SAGO – THE EQUATORIAL SWAMP AS A NATURAL RESOURCE

WORLD CROPS:
PRODUCTION, UTILIZATION, AND DESCRIPTION

VOLUME 1

Also in this series

Volume 2
Improvement of Quality Traits of Maize for Grain and Silage Use. A seminar in
the EEC Programme of Coordination of Research on Plant Protein Improvement
held at Schwäbisch Hall, Federal Republic of Germany, November 29 - December 1,
1978
edited by W.G. Pollmer and R.H. Phipps

SAGO publications

Executive editor: TAN Koonlin
Editorial assistant: Irene NG Bee Chin

SAGO-76: Papers of the First International Sago Symposium, edited by TAN Koonlin,
is available postpaid at £12 sterling or US$25.00 (bank draft) from:

KEMIKRO Sdn. Bhd.
Petaling, P.O.Box 46
Old Klang Road
Kuala Lumpur
Malaysia

Enquiries concerning SAGO-76 should be addressed to·

Dr. TAN Koonlin
KEMIKRO Sdn. Bhd.
Petaling, P.O.Box 46
Old Klang Road
Kuala Lumpur
Malaysia

SAGO
The Equatorial Swamp as a Natural Resource

Proceedings of the Second International Sago Symposium, held in Kuala Lumpur, Malaysia, September 15-17, 1979

edited by

W.R. STANTON and M. FLACH
Kuala Lumpur, Malaysia

1980

MARTINUS NIJHOFF PUBLISHERS
THE HAGUE / BOSTON / LONDON

Distributors:

for the United States and Canada
Kluwer Boston, Inc.
190 Old Derby Street
Hingham, MA 02043
USA

for all other countries
Kluwer Academic Publishers Group
Distribution Center
P.O.Box 322
3300 AH Dordrecht
The Netherlands

This volume is listed in the Library of Congress Cataloging in Publication Data

ISBN-13:978-94-009-8930-6   e-ISBN-13:978-94-009-8928-3
DOI: 10.1007/978-94-009-8928-3

CONTENTS

MENTERI PERTAHANAN MALAYSIA

M E S S A G E

I have followed the sago saga, from its inception, with keen interest. As Minister of Information I was happy to encourage research and awareness of the plant through the making of the film "Pohon Bertuah" which depicted the traditional way in which sago was part of the fabric of life of the people of the Region, particularly that of my own community the Melanau.

I appreciate the need to encourage continued effort in the development of diversity in the supply of basic local resources, as part of the strategy of political insurance against the hazards of the world of tomorrow. I welcome the Conference Sago '79, as a contribution to the study of stability in the Region.

(DATO' AMAR HAJI ABDUL TAIB BIN MAHMUD)

MINISTER OF DEFENCE MALAYSIA

Jalan Padang Tembak,
KUALA LUMPUR 15-03

Telefon: 24287
87731 Samb. 208

Kawat: KEMENTAH
KUALA LUMPUR

Tal. Pejabat No.: K.L. 87568
Alamat Kawat: "KEMTEK
         KUALA LUMPUR"

MENTERI SAINS, TEKNOLOJI DAN ALAM SEKITAR, MALAYSIA
(MINISTER OF SCIENCE, TECHNOLOGY AND ENVIRONMENT, MALAYSIA)

I am happy to send my congratulations and best wishes to the organizers of the second International Conference on Sago to be held in Kuala Lumpur from 17th - 19th September, 1979. I am indeed honoured to have been invited to officiate at the Conference.

Since the first International Conference on Sago in 1976, the organizers have continued to sustain interest in regard to Sago and have compiled the proceedings of the last conference into a book. The knowledge contributed by local and oversea participants and experts is an invaluable source of co-ordinated information and there is no doubt this second International Conference will bring about more information regarding the Sago industry.

I would like to join the organizers in welcoming both the local and foreign participants and experts to this Conference to contribute their knowledge from their respective countries and regions. As for me I personally have a long standing interest in the Sago industry having come from Sarawak which has vast areas of peat swampland and where Sago has been an important industry in sustaining the native communities in the earlier days. As such I hope this Conference would create and arouse the necessary interest both locally and overseas to revive the Sago industry, to make it an important food crop and to develop it into an economically viable concern by utilizing the vast swamplands of Malaysia.

I am informed that this Second Conference would focus attention on the management and utilization of Sago as an important industrial raw material and also to explore Sago as a convenient form of energy, which could be tapped for the benefit of mankind.

I shall look forward to the findings of the Conference and to wish it every success.

(Tan Sri Ong Kee Hui, PMN., PNBS., PGDK.)
Minister of Science, Technology & Environment,

Kuala Lumpur, 1st September 1979.

DEPUTY MINISTER
TRADE AND INDUSTRY
MALAYSIA

Talipon: 946589
Taligeram: DAGANG, KUALA LUMPUR

TIMBALAN MENTERI
PERDAGANGAN DAN PERINDUSTRIAN
MALAYSIA
WISMA DAMANSARA
KUALA LUMPUR 23-03

27th August, 1979.

M E S S A G E

      I deem it a great honour and privilege to send you this message on the occasion of the Seminar on SAGO products.

      Southeast Asia is a region rich in renewable resources. We, who are planning for the future, recognise that in the past the major products of our lands have been exported in an unprocessed state. Further, the range of major renewable resources exported, especially from Malaysia, has been comparatively limited. The region has also imported in quantity, processed plant commodities which are capable of being used for manufacturing locally.

      I therefore welcome this timely Seminar on SAGO and other crops of the equatorial swamp lands as a help to the effort of the Ministry of Trade and Industry in its encouragement of manufacturing based on local raw materials. For, I find difficulty in identifying a plant feedstock for industry possessing greater diversity than that of starch. If, as the organisers suggest from their introductory notes, we can pursue industrialisation without detriment to the environment, then the Seminar will be doubly welcome.

      Although I will not be able to be present in person due to other commitments, I am happy to send greetings to all the participants, especially those from our fellow ASEAN countries.

(Dato' Lew Sip Hon)
Dy.Minister of Trade and Industry,
Malaysia.

Sime Darby Berhad
Wisma MISC, 6th Floor, Jalan Conlay, Kuala Lumpur 04-09
Tel Nos. 423939 & 426044
Telex: SDMAL 30038
Cable: "SIMDARB KL"

Opening Address

I must admit that when I was approached to preside at the opening ceremony of this Symposium, I wondered why it was being held at all, as sago has always been associated in my mind with a dessert called sago pudding, of which I have never been particularly enamoured, apart from the "gula melaka" with which it is usually served. Apparently, things have changed quite a bit since then.

Our scientists feel that, apart from its value as a food, sago has other uses which are generally not known. In particular, it could be a source of fuel and this is clearly an important consideration in view of the impending worldwide shortage of energy. Another advantage of sago cultivation is that land planted with sago does not require extensive engineering or costly maintenance. It generates less waste than some other crops since all the parts are usable. I am also told that it conserves the climate, soil and wildlife.

The big question mark is its profitability and general economic viability compared to the other plantation crops which are now being planted, i.e. rubber, oil palms and cocoa. If anything, this Symposium should explore the desirability of diversifying our plantation industry. In this respect, we cannot afford to stand still. As the saying goes, we should not put all our eggs into one basket. Of course, at the moment, we have put them into more baskets than we did 40 years ago, but as a matter of principle, we should diversify even more if such a course of action is economically viable.

If I may say so, this should be the main theme of this Symposium, apart from the technical arguments which will be brought forward in the course of the discussions. They should certainly be revealing and point a path towards the future.

Tun TAN served in the Government of Malaysia, inter alia, as Minister of Finance during 1959-74.

Tun **Tan Siew Sin** (SSM, JP)
CHAIRMAN

MENTERI PERTANIAN MALAYSIA

Opening Speech by Dato' Shariff Ahmad,
Minister of Agriculture, Malaysia
at the Opening of
the Second International Sago Symposium
Kuala Lumpur, 17th September, 1979

I feel greatly honoured to be invited to declare open SAGO-79. I would like to thank the organisers and to commend them for their effort in organising this Symposium. I understand that this is the Second International Symposium on Sago held in this country, and the previous symposium was held in Kuching, Sarawak in July 1976.

The theme of this Second Symposium, "Starch Power to Beat the Fuel Crisis and the Equatorial Swamp as a Renewable Resource" is very appropriate in these times of energy crisis and escalating fuel prices. I am sure all of us welcome any effort made to explore the efficient use of sources of energy other than petroleum; and I urge all the experts to look realistically at this possibility. I am also happy to note that the question of cost of production, method of processing, and the increase in the range of end uses of starch will also be discussed in the Symposium.

The fuel shortage and price escalations have given me a lot of worry because they will affect our agricultural mechanisation programme. It has been our declared policy to mechanise agriculture in this country as a lever not only to increase food production but also as an incentive to youngsters to take up agricultural farming. Our effort to mechanise agricultural farming is one of the most important objectives as Malaysia is basically an agricultural country and about 60% of our population are dependent upon agriculture for sources of income, but their income is still below our required target.

As the industrialisation programme in our country has gone very far ahead, using modern equipment and automation which has attracted the young-sters a great deal, the tendency to leave the rural agricultural environment for employment in the urban areas is increasingly widespread. To me, the answer to this lack of interest in youngsters is to modernise our farming techniques.

We have started quite well in this direction, but the oil shortage and price increases are causing me a lot of concern. The Ministry of Agriculture alone consumes about 5 million gallons of diesel every month, so as to enable farmers to use tractors in the field, and fishermen to modernise their fishing equipment, such as outward boat motors, etc. I am confident the Symposium will examine and successfully find the best solutions to beat the fuel crisis; if not today, perhaps in the not too distant future.

In the past, emphasis and attention had been given to two principal plantation crops, i.e. rubber and oil palm, which are grown on a very large scale. This was, to a great extent, due to their revenue earning capacity. Sago, however, has not been of great importance in Malaysian agriculture except for the State of Sarawak. The exploitation of the sago palm is on a very small scale in Peninsular Malaysia and in Sabah. However it is of high significance both in nature and magnitude in Sarawak. Although revenue from sago is no longer the mainstay of the economy of Sarawak, it remains a very important agricultural export of the State as well as providing employment to many thousand families, especially in the Third Division.

The total area under sago in Malaysia was 18,713 hectares in 1976 with 2,963 hectares in Peninsular Malaysia and 15,750 hectares in Sarawak. The average annual tonnage of sago flour exported is about 29,800 tons over the last five years, with Sarawak accounting for about 26,500 tons and Peninsular Malaysia for about 3,300 tons. Britain was the main importer of our sago in the early 1950's. However, Japan has currently assumed that position and alone accounted for 68 per cent of all sago export from Malaysia in the last five years.

The sago palm has a number of uses to suit the needs of changing times. In the past, the starch extracted from the sago palm once formed an important source of food for the indigenous people, and the leaves of the palm were made into superior quality attap roofing. Today, the main

uses of sago starch are in the form of sweeteners, thickeners, confectioneries
in food manufacturing, textile finishes, jelling agents and as animal food
component. I am sure, in the near future, it could economically and
effectively be used as a source of fuel; and, of course, it needs a lot
of research work to be done before we can arrive at that formula.

As you all know, the sago palm thrives well in peat areas. Such
areas require extensive engineering and costly maintenance to exploit for
other plantation and annual crops. Since Malaysia has an extensive area
of about 2.3 million hectares (5.6 million acres) under peat, the potential
for sago cultivation is tremendous. This is particularly so in the State
of Sarawak where extensive peat areas cover about 1.5 million hectares
(3.6 million acres) or some 12 per cent of the land area in the State.
Pending technological advancements and the requisite infrastructural
development that would pave the way for other economic crops, I feel that
the development of sago would provide a solution for those peat areas
where no alternative crops seem feasible.

I must admit that at the moment very little research has been carried
out on the sago crop in this country. Thus there is a need at least in
Sarawak for a comprehensive research programme to be undertaken, and I
am directing the Malaysian Agricultural Research Institute (MARDI) to
help sustain the industry. The research programme should include:-

(1) Investigation into the development of better methods and
    techniques in the cultivation of the crop; and
(2) Investigations into the production of better varieties of sago
    palms which give higher yields of palms per acre and higher
    starch per palm.

I have been given to understand that this Symposium is attended by
experts from our ASEAN countries as well as from Japan and Europe. There
can be no doubt that the exchange of ideas, thoughts and experiences
during the next two days will be of great benefit to all concerned. The
meeting of minds across national borders and international waters can
only result in ideas that will bring good to the world as a whole, and
especially to new developing nations that are so badly in need of new
technologies and support in their efforts to upgrade their socio-economic
well-being in line with international standards. Symposiums such as
this also perpetuate the spirit of international co-operation and help
to fulfil the need to reduce gaps among nations.

We in Malaysia are always open and receptive to new ideas, concepts and thinking on any subject pertaining to our development efforts in general. We also welcome new and constructive ideas that are intended for the good of all. Thus, I believe, your deliberations may raise issues and solutions which may be very relevant to us. It is on this basis that I look forward to the result of your discussions and deliberations because I firmly believe that there is much that you can contribute to the development of the sago industry in our country.

Finally, I would like to extend a most hearty welcome to our foreign visitors to Malaysia and hope that, apart from working hard at the Symposium, you will find time to look around our country.

It now gives me great pleasure to declare open the SAGO-79 Symposium and I wish you all every success in your deliberations.

(DATO' SHARIFF AHMAD MP)

The Plantation Sagopalm,
Metroxylon sagu Rottboll

SAGO-79

Recommendations

STARCH PALMS AND THEIR ENVIRONMENT

The Symposium noted the valuable role starch and sugar palms, not only the swamp sago, Metroxylon sagu, but also others which inhabit more upland and savanna country, had played in the past in the sustenance and development of the people of insular Southeast Asia. It recommended that fundamental studies should be made on the genetic variability and productivity of the different palms, as a basis for their future exploitation for industrial and food uses.

The soundest approach would be, in the absence of experimental data, to use the inherent knowledge of the indigenous people, currently the only real "sago experts". The folklore is of value in aiding scientific identification of the properties of different palms for formulation of the research and development programme. As societies in which sago has been used have dwindled rapidly, it is recommended that a comprehensive record should be assembled of this traditional knowledge on the most useful clones and how they are identified, the cultural practices, the types of soils, sterile palms, pests and their control, and other practices relevant to the cultivation of particular species.

It was noted that there was still a lack of precise knowledge of the extent and character of the swamplands of the Region and their suitability for the establishment of sago palmeries. It was therefore recommended that study of the palm environments should make use of currently available aids to terrain surveying, e.g. remote sensing, aerial and satellite photography, to locate suitable areas that should be reserved for the development of diversified carbohydrate palm based industries.

W.R. Stanton and M. Flach (eds.), SAGO. The Equatorial Swamp as a Natural
Resource. Proceedings of the Second International Sago Symposium. All rights reserved.
Copyright © 1980 Martinus Nijhoff Publishers, The Hague/Boston/London.

The Symposium stressed the need for creating general awareness, but especially amongst land-use planners, that the palms
are economic plant species adapted to specialised ecosystems,
for the maintenance of which there was a need to preserve plant
cover in depth. This conservation was in the interests of the
long term environmental welfare of the Region; thus the environment in which the present wild palmeries thrive needed to be
characterised in detail.

It was noted that the quantitative assessment of solar
radiation was not available throughout the Region for calculating
palm productivity and energy input. It was recommended that
this statistic should be incorporated in the inventory, noting
in addition the quality of the radiation. The value of the latter
information had been shown, for example in Sweden, for its application to harnessing plant growth for biomass production at
comparatively low levels of radiation.

ECOLOGICAL ADAPTATION

It was noted that the swamp palms had physiological, anatomical and reproductive features associated with their specialised environments; similar plant features had also been noted
in the cold swamp environments in Sweden. The Symposium recommended that there should be interchange of information between
plant-biomass research workers in the two environments, and that
the techniques for pollarded plantings of lignocellulose-producing
trees should be considered alongside the study of starch-producing
trees, the palms.

The Symposium recommended that note should be taken of
current studies on the mangrove environment, another highly
specialised plant biomass producing system of value per se, viz.
without alteration of the water regime; it was noted that mangrove species also served as staple food plants in parts of the
Region. It was recognised that the boundary between these two
ecosystems was adjustable.

It was also noted that, notwithstanding its current exclusive
association with dryland agriculture in Malaysia, experience
with the oilpalm, also a swamp denizen in its original home in

West Africa, suggested the possibility of selecting and breeding
a less hydrophilic sagopalm for extension on dry or non-inundated
habitats. Further, as with the Dumpy oilpalm, a dwarfer sagopalm,
bred also for earlier maturity, would be a more manageable crop
that would allow a higher planting density and quicker returns.

RURAL DEVELOPMENT AND PLANTATION AGRICULTURE

It was noted that, for the development of a carbohydrate-
based industry, large scale planting would be required in order
to achieve adequate supplies of feedstock for installations of
optimum capacity. The Symposium recommended however that the
strategy of land use currently being adopted in Papua New Guinea
(as reported at the Symposium) should be studied by other
countries. A particular social feature was that of encouragement
of participation by individual farmers. This practice conserved
the agrarian form of the local society and avoided making a few
individuals rich at the expense of the community. This method
of dispersed rural industrialisation was generally regarded as
appropriate to the tropical countries.

It was noted that highly successful plantation treecrop
monocultures had already evolved in several parts of the Region,
which had experience of the industrial infrastructure required
for such agricultural enterprise. Malaysia, in the states of
Sarawak and Johor, had developed a higher level of expertise in
commercial palm-starch arboriculture than anywhere else in the
world.

Governments were recommended to note and study the implications of the shift of energy sources towards renewable fuels,
viz. the changes in the location of population and industry
arising from plantation-based energy and feedstock.

AGRONOMY

It was noted that the present state of the carbohydrate
industry was analogous to that of the oilpalm industry in Africa
at the beginning of this century, before the plantation oilpalms
were first cultivated in Southeast Asia, viz. plantation industrialisation of the starch palm must be initiated on a primitive

cultigen. Whilst evolution under domesticity is essential, note was taken of the danger of loss of diversity which might arise, and therefore the need for conservation of the primitive stocks.

The Symposium recommended that full advantage should be taken of studies on other palms such as the oilpalm, coconut and datepalm, for which crops more reliable information on biology and commercial management was available. It was noted however, that the carbohydrate palms were generally freely tillering, and the Symposium recommended study of this phenomenon in the establishment of élite clones as well as the new method of tissue culture.

It was noted that, through selection of open-pollinated seed, thornless sago clones had arisen from spinous palms; these thornless cultivars may need crossing with other disparate stock to reinforce the positive features of the palm. The Symposium recommended that, in addition to conservation of wild stocks, suitable clones of all potentially useful species be collected for the gene pool, to facilitate studies on the comparative rates of starch accumulation during development in order to determine earlier harvesting time of particular clones.

The fact, observed by farmers in the easterly parts of the archipelago, that sterility might promote high yield in the sago-palm was further noted. It recommended, firstly, the study of induced sterility via suppression of flowering (or removal of flower initials), to examine the commercial value of this tech-nique for increasing productivity; and secondly, the nature of genetic and environmental influences on palm maturity because, under aboriginal exploitation, selection of productive sagopalms had largely ignored a feature that has economic implications for extensive commercial exploitation, i.e. the duration of the starch accumulating period of growth.

It recommended that dangers from diseases and pests, inherent in large scale monoculture, should be taken into account, and that in the research for productivity the techniques of poly-cultivar monocropping, as is being adopted in Sweden for trees and worldwide for cereals, should be considered.

PRODUCTION TECHNOLOGY

The Symposium recommended that in the pursuit of development, care should be taken with choice of machinery not to disrupt the local social system and edaphic conditions. It was noted that after logging, palm trunks were easily preserved by total immersion for a relatively long period without significant loss of carbohydrate.

The Symposium recommended that special attention be given to the study of methods of harvesting, transport and pond storage which would avoid biodeterioration, e.g. storage of whole trunks is preferable to hauling short logs necessitated by the current method of handrolling for assembling of rafts. Machinery used in timberlogging could be modified for handling sagologs, such as debarking and feeding into the grating machine mechanically, so as to achieve higher input of raw materials, especially where there is a shortage of labour.

CARBOHYDRATE EXTRACTION

It was noted that countries differed in their preference for the end-use of carbohydrate. Because of their different socioeconomic conditions, the Indonesians stressed the importance of food carbohydrates, whereas Papua New Guinea had opted for liquid fuel production.

It was noted that most sago-starch, as at present extracted, is less useful for subsequent processing than tapioca-starch, the best being only "acceptable" but not even "comparable", partly because of high fibre content; sago-starch was also said to be more difficult to deodorise. The Symposium therefore recommended that the factors in biodeterioration associated with the several unit processes of harvesting and extracting the starch should be identified and their control investigated.

It was noted that mechanisation in the factory would be more easily and profitably achieved than in the field, in view of the preferred habitat of the palm. It was noted that in the last decade, there has been significant technical advances in tropical starch processing, in particular the adoption of centrifuging in tapioca-starch extraction in Peninsular Malaysia.

Enlarging the scale of commercial sago processing would support
a higher level of technology in both processing and manufacturing
to make sago-starch more competitive with other starches, sugar
and feedstuff materials.

It was noted that, given a free market, the demand for
starches was highly flexible and that the factors affecting use
and trading in these commodities were diverse, the starch market
being also affected by other markets, chiefly mineral fuel, sugar,
cereal and other animal feedstuff. It was noted that this market
information was required by both producing and importing countries. The Symposium recommended that assistance be sought
internationally to produce a comprehensive basic study on the
international starch market, as it affected palm starches and
sugar, for use by individual governments and specific plantation
schemes.

The diversity of possible downstream products from the
postulated carbohydrate palm industry was noted. But the criteria
for the preparation of the feedstock for power alcohol differed
from those for starch production, and the Symposium recommended
that research should be conducted to study the optimising of
the generation and extraction of "fermentable" carbohydrates,
as feedstock for palm product based fermentation industries.

FOOD PRODUCTION

It was noted that there was current neglect of the carbohydrate palms as resources for food in the Region, which was
attributed to subsidies to more prestigious foods, particularly
rice in Indonesia, to a socially acceptable level that was
straining scarce capital reserves. As an aspect of the palm
resources survey, the Symposium recommended that a regionally
standardised method of surveying plant food and fuel resources
be adopted, taking note of the hidden social costs of subsidising
specific crops and commodities. This feature naturally was
stressed by Indonesia, but it was noted as applicable to other
tropical countries.

It was noted that it was energy food not protein that was
the critical food scarcity in many tropical areas. This could

be solved in relation to the high potential food yield produced from palm carbohydrates per unit area and, because of the high nett-energy gain in farming of these crops, governments should resuscitate their husbandary. In the light of future food needs, the Symposium recommended that efforts be made to raise the current low status accorded to indigenous starch palms within the foodcrop hierarchy.

The value of the carbohydrate palms in increasing the efficiency of distribution and food use of the available protein was noted, and it was recommended that two technologies be given serious consideration:

(a) The use of palm starch and sugar in food-industry, i.e. fermentation, meat and fish processing, production of amino-acids, etc., and

(b) The use of traditional sago products with good keeping quality, such as briquettes, to increase the available food reserves, especially for use as an emergency or relief food.

It was noted that in satisfying human food needs, there were cheaper ways of increasing protein food supplies than by the genetic manipulation or cultivation of protein-rich but low yielding crops, e.g. by fortification (using part of the carbohydrate harvest to synthesise the additive), as well as by in situ enrichment.

Currently very few alternative crops had been found to be commercially viable in the equatorial swamp habitat without adopting costly and, in the long term, environmentally and pedologically unsatisfactory cropping and land management practices. This was recognised as one of the causes of abandonment of so-called reclaimed swampland.

The Symposium expressed concern over the extent of abandoned, misused and underused land, such as neglected sawahs, throughout the Region. It recommended that an inventory of the areas be made for rehabilitation to a productive state as part of the palm culture, although it was recognised that the traditional farmer too was profit-motivated, and realistic consideration must also be given to the alternative uses of the swampland. WRS

The Original Grainless Hoe Culture of Southeast Asia

A floodplain dweller in Ramai village, Ramu estuary, cutting out the pith of the sagopalm, to make a flour that is cooked into a gelatinous gruel, which forms his staple diet for much of the year.
Papua New Guinea

Illustration courtesy of CSIRO, Canberra

ECOLOGICAL NOTES ON SAGO IN NEW GUINEA

K. PAIJMANS

Sagopalm, Metroxylon sagu, is found over hundreds of square
kilometres of floodplain swamp in New Guinea, and covers many
small areas in beach plain swales and swampy mountain valleys.
It occurs naturally from sea level to approximately 1000 m
altitude, but has probably spread widely beyond its natural
range through planting. Sago grows on silt, clay and peat of
alluvial plains and backswamps as well as on sand in some coastal
localities, and on sites that are fresh or slightly brackish.

The palm grows best in shallow swamps where there is a
regular inflow of fresh water, such as inlet zones of swamp
margins, and where the water table is below the surface outside
the flooding season. On such sites its fronds are up to 14 m
high, and the flowering starch-producing tree-like stems reach
over 25 m. The palm multiplies, commonly after flowering, by
growing suckers around the base of the old stems, thus forming
stools of 3-5 palms. Both smooth and spiny varieties are present,
but the variety with needle-like spines along the frond midrib
is more common than the unarmed type.

All gradations occur from stands of pure sago virtually
without trees to woodland with a rather dense layer of trees
and an open lower tier of sago. Together with species of
Pandanus, sago commonly forms an open understorey in swamp
forests of Campnosperma spp. and Terminalia brassii, and occasionally also under Melaleuca. Melaleuca swamp forests, flooded
by 1.5 m or more of water in the rainy season, largely dry out
in the dry season, when trees and undergrowth, including sago,
are often severely damaged by fires lit in nearby grassland
and spreading into the forest. Other associated trees, all
tolerant of wet conditions, include Alstonia scholaris, Bischofia

Forested Flat Fan Surface with Sago Vegetation in Swampy Depressions of Intermontane Lowlands in Vanimo Area

javanica, Nauclea coadunata, Planchonia papuana, and species
of Syzygium and Neonauclea. On slightly brackish coastal sites
the stilt-rooted Myristica hollrungii is a frequent companion.

The ground layer of a sagopalm stand varies with the density
of the palms and the degree of swampiness. In dense well-
developed stands the long fronds, overarching from the base,
form a closed canopy, the interior of the stand is gloomy, and
there is no undergrowth. The peaty waterlogged soil is layered
with fallen dead fronds, and the palm's numerous pneumatophores
form the only live "cover". Open stands have a ground layer
of grasses, gingers and ferns where the water table is below
ground level at least for part of the year, and an undergrowth
of shrub pandans, tall coarse sedges, Hanguana malayana or
Phragmites karka where the water table is permanently at or
above the surface.

Sago stands become permanently stunted in transitions to
deep herbaceous swamp, in brackish environments, and on sites
where the water table temporarily sinks deep enough to cause
drought stress. On such sites the palms do not flower but
sucker strongly and may grow quite densely. In transitions to
permanent herbaceous swamp and reed swamp, sago grows in scattered rounded groves which are up to 100 m across.

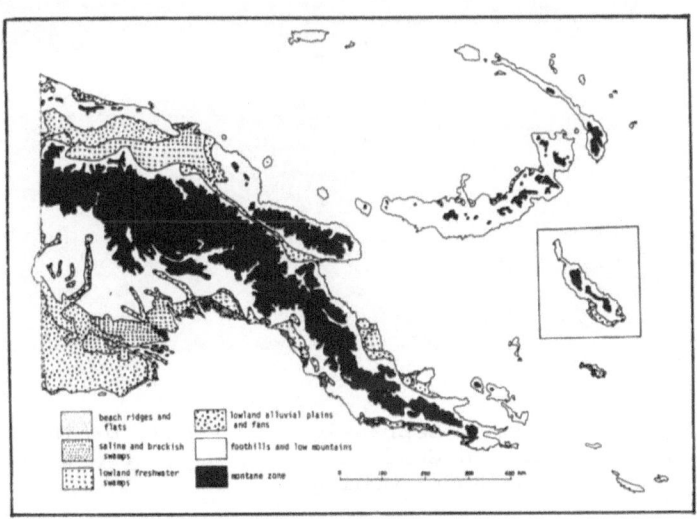

Physical Regions of Papua New Guinea

Sago and Remnant Forest in Valleys between Low Hill Ridges,
common in the Sepik area

Sago Understorey in Swampy Depressions on Poorly Drained
Forested Alluvial Plain in Gulf District, subjected to wet
season flooding

LOGGING THE SWAMP FOR FOOD

TAN Koonlin

USEFUL PALMS IN SOUTHEAST ASIA

Palms abound in Asia more than in any other continent, in a diversity of warm habitats, with the richest collection within the equatorial belt; the few species which persist in less tropical latitudes mark the periphery of an extensive realm when the earth throve in a warmer climate. Perhaps more than any other, the heartland of this family of plants has remained intact since the Cretaceous, for it is confined to that part of the world where climatic changes were least extreme, 10°N and S.

The notable palms which are exploited for subsistence or commerce in Southeast Asia provide four types of products:

(a) The woody palms: Calamus spp. and other rattans, and the nibong, Oncosperma filamentosa, and the pinang, Areca catechu, whose durable trunks are commonly used for housing and fishing structures;

(b) The oleagineous palms: chiefly the ubiquitous coconut, Cocos nucifera, and its more productive rival, the African oilpalm, Elaeis guineensis;

(c) The sacchariferous palms: the nipa, Nypa fruticans, whose sap may be the only potable liquid available in some brackish areas (Appendix), the fishtail-palm, Caryota urens, the aren or sugarpalm, Arenga pinnata, the lontar, Borassus sundaicus, and its more northerly kin, the palmyra, B. flabellifer, which favour savanna-type habitats; and

(d) The farinaceous palms: the swamp sago, Metroxylon sagu, and its wild precursor, M. rumphii, the gebang, Corypha elata, and hill-sago, the panto, Eugeissona utilis, and the kajatto, E. insignis, which colonise upland slopes; some of the sugar palms may also be used for sago as famine food.

14

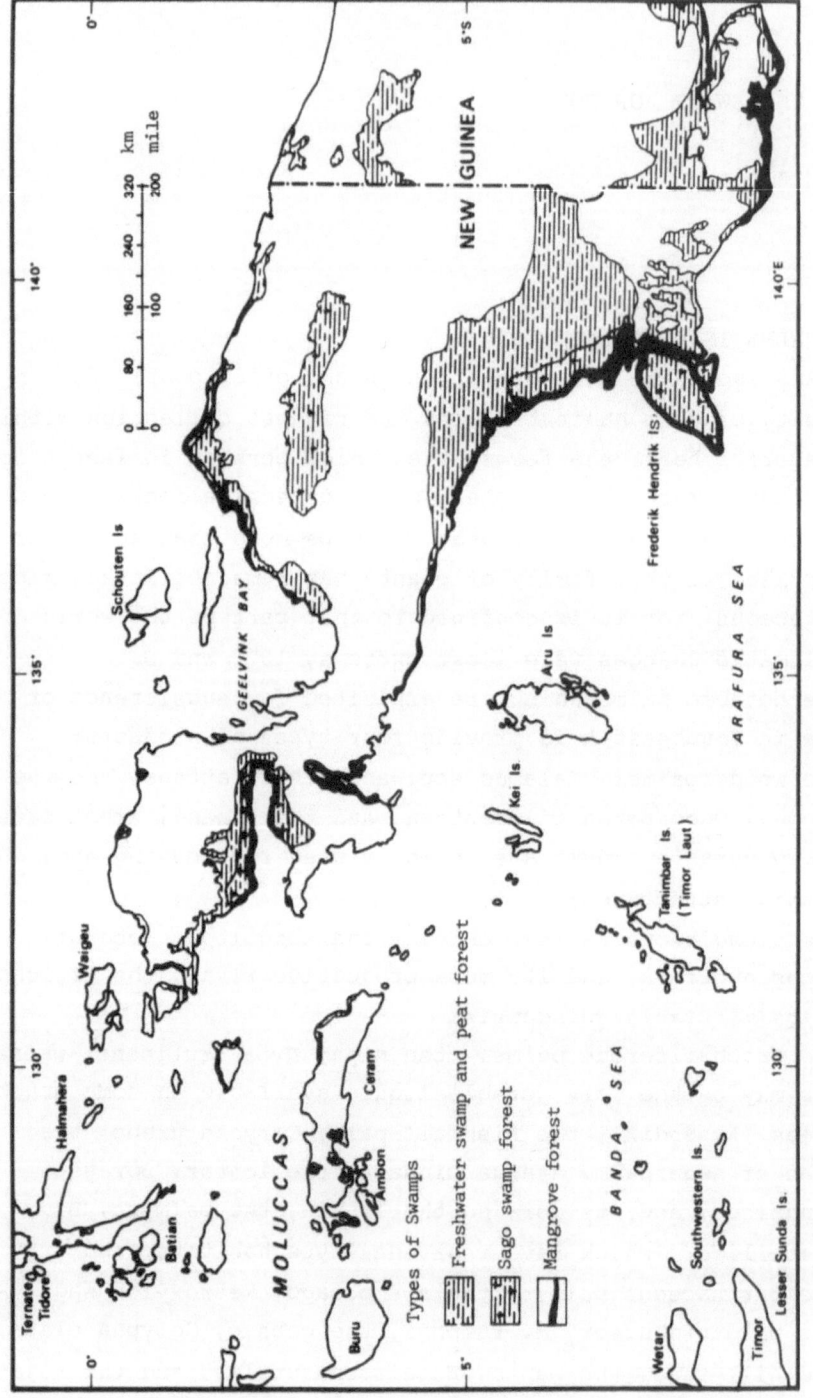

Figure 1. Swamps of Eastern Indonesia

Based on Vegetation Map of Malaysia, comp. C.G.G.J. van Steenis 1958;
UNESCO Humid Tropics Res. Proj., 1 : 5,000,000.

Palms are perennials that when cropped yield a bounty in shelter, food and drink that has given rise to subsistence strategies which are distinct from those based on the better-known annuals, the grains and tubers. Several cultures have developed a largely self-sufficient domestic economy based on palm arboriculture, e.g. swamp Amerindians of the Amazon-Orinoco on the moriche, Mauritia flexuosa, the Arabs of sub-Sahara on the date-palm, Phoenix dactylifera, the South Indians on the palmyra, the West Africans on the oilpalm, the Indo-Pacific Islanders on the coconut, the Moluccans and Papuans on the sago (Burkill 1935), and the non-eating people of the Lesser Sunda Islands (Nusa Tenggara) on the lontar drink (Fox 1977); unique are the Tasaday of Mindanao, a Stone Age group which acquired its sago staple, from a Caryota sp., in the 1960s.

Of the sagopalm, Logan (1849) described its natural setting most clearly:

> "In most parts of the Indian Archipelago two kinds of alluvial soil are found in greater or less abundance, one consisting chiefly of sand often thrown up in long banks, and the other chiefly of decomposed vegetable matter ..... For these two descriptions of soil, nature has provided two kinds of palm adapted in a wonderful manner to the necessities of man. On the barren sand she has planted the coconut, and in the morass the sago tree".

But while the importance of the coconut in indigenous economies in insular Southeast Asia is wellknown, that of the carbohydrate-rich palms has become unrecognised in the modern era, when in fact of the "Papuan food-plants, the sago-palm, long known from there, holds a rank disputed only by the Cocos-palm" (Thomson 1892).

EQUATORIAL SWAMP SETTLEMENT

The Metroxylon palms are a singularly insular genus: the recognised species are confined to the humid areas of and endemic on New Guinea and neighbouring islands (Figure 1). M. rumphii is commonly found in Maluku and New Guinea; towards the west the cultivated spineless form, M. sagu, prevails, which through selection seems to have also evolved into a more massive and longer maturing palm. The loss of a characteristic primitive palm trait

most recently may be attributed to the Mohamedanised cultures
of the western archipelago which do not harbour the pig, especially from the 14th century, and hence the armoured palm generally
became superfluous; selection for palms that could be easily
handled for building materials probably reinforced the progress
to spineless forms even in areas where the palm was not a food
source. This distribution of the two species recalls the faunal
dichotomy of the Wallace Line.

Selection of productive cultivars has entirely been achieved
empirically, under aboriginal agronomy, comprehensive scientific
breeding still not instituted by the late 1970s anywhere in the
world. Genetic studies on the cultivars in the westerly islands
may throw light on the domestication of the species. Although
it seeds itself, the common mode of propagation under regular
exploitation is vegetative, partly in the belief that seeding is
"generally unproductive" (Low 1848); Burkill (1935) noted that
sago seeds in Malaya and Borneo too were commonly infertile.
Further, seedling palms are of variable quality. Logan (1849)
pointed out that regeneration "by radical shoots, exactly in the
same manner as the common cultivated plantains, is peculiar and
is not observed in the true palms". Cantley (1886) reported the
period of immaturity to be six years for seedling palms, and
Forrest (1780) seven years for natural stands, but modern writers
incline to attribute a much longer vegetative phase to this palm.

Unlike other farinaceous palms, the sagopalm is a hydrophyte:
"a good sago plantation or forest is a bog knee-deep" (Crawfurd
1820), "low marshy situations shut out, but at no great distance
from the sea, and well watered by freshwater seem most productive"
(Logan 1849). In a conducive habitat in New Guinea, palm fronds
are up to 14 m high and flowering stems reach 20 m. Lush groves
may be found on lowlying swampy swales behind beach ridges, along
river banks, lake margins, creeks, pools, gullies, even "swampy
hollows on the rocky slopes of hills, where it seems to thrive
equally well as when exposed to the influx of salt or brackish
water kept constantly full of moisture by the rains, and by the
abundance of rills which trickle down among them" (Wallace 1898),
up to altitudes of 2,500 ft, e.g. on Manus Is., Lake Kutubu and

the Torricelli Mountains, New Guinea (Ryan 1972).

"The sago and mangrove swamp areas are inimical to human life ... furnished with conditions which to European life would be most inhospitable, if not almost altogether insufferable ... unfit for habitation" (Thomson 1892). But to the native, "the flora of the wetlands, the swamp, is far more important in almost every way than that of the hillside undergrowth of the rainforest ..... As a source of produce requiring no kind of care, the swamp is a mudmine of lasting merit" (Harrisson 1970). Papuan sago eaters even dam up water courses to create swamps in which to nurture their palms, and the Bisaya used the notable swamp denizen, the water-buffalo, to log their sagoswamps, pulling logs and boat-loads of crude sago overland.

Maher (1961) described the symbiotic relationship man had developed with perhaps the most productive swamp he chose to inhabit: "A physical environment of rivers and sago swamps ... offers enough food for those who know how to get it and efficient transportation for those who know how to use it, but it also is poor land for walking or farming ..... An important combination of environment and technology ... allowed broader limits or more alternatives for their collecting activities ..... with the rivers and the easy productiveness of the sago palm an alternative of large, concentrated communities was added". "The dense growth of sago served as bases for victually for the crews of the prahus, enabling them to roam for long periods outside their own village" (Held 1957), and "commercial relations in export trade with neigh-bouring tribes are fostered" (Thomson 1897). Aboriginal groups circumvented the mosquito menace by dwelling in brackish fringes and migrating regularly to "hotel" villages in the sagoswamps to harvest the palm produce.

The swamps in New Guinea and Maluku have stimulated no comparable extent of sedentary cultures dependent on an indigenous domesticate other than the aroids, which are often unreliable food staples, nor do they store well, and the sagopalm not infre-quently is retained as a supplementary or seasonal staple because of its hardiness and abundance; not uncommonly, marginal humid areas become habitable because of the presence of this palm.

In the Aitape-Ambunti region (4,650 sq miles), "subsistence cultivation ... virtually everywhere ... is supplemented by sago collecting ..... on the island plains and Sepik flood-plain the situation is reversed, and subsistence cultivation is subsidiary to sago collecting" (Haantjens et al. 1972); generally wherever sago thrives, "the population concentration ... seems to be due mainly to the abundance of sago ..... Everywhere sago is an important part of the diet" (Löffler et al. 1972).

That sagoland can also have a high population capacity is apparent in the New Guinea lowlands. "Population density ... per sq mile of 100 to 250 ... is largely due to a great reliance on sago and/or fishing for subsistence. In the hills and mountains there is a more general tendency for high population densities to be associated with intensive use of land for cultivation; nevertheless, even here the high population densities are partly sustained by reliance on sago" (Haantjens et al. 1972); "within the Maprik area ... densities on tribal lands ... reach over 400 persons per sq mile, the highest recorded in the lowlands of New Guinea" (Haantjens et al. 1968). The sago complex manifests an impressive spectrum of technical skills necessary to fell large trees and extract flour from their pithy stems which can be made into staple foodstuffs, that belies its image of cultural backwardness.

SAGO ARBORICULTURE VS WILD PALMERIES

Many colonial observers who wrote at some length on the sago-palm noted in effect what Crawfurd (1820) described: "Of all the plants which afford a supply of nutritious farina for human aliment, the sago affords at once the most obvious, easy and abundant one". This flour, when made into a toast, could keep for a dozen years, which Forrest likened to "hot rolls ... equal to our wheat bread", and the Malay sailors despised as roti papua.

Wallace estimated that a palm trunk 20 ft long and 4-5 ft circumference could feed a man for a year, producing 900 lb of wet flour that would bake into 600 lb of bread, 3 slices to the pound, his daily ration being 5 slices. This harvest took the equivalent of 10 mandays for reducing the pith to an edible meal,

and 10 womandays to bake the bread in the "papua oven", prototype
of the modern toaster. Forrest (1780), who frequently supervised
his crew in making sago bread to victual his ship, put the annual
subsistence of a man at three palms, each palm producing 3 cwt of
flour. Given a 7-year immaturity, one seventh of an acre with 300
palms could be logged annually to sustain 14 men, and five men's
labour would feed 100.

Crawfurd's (1820) estimate was 8,000 lb of dry flour per acre,
each palm yielding 500-600 lb. Logan (1849) calculated that the
palm, yielding 700 lb of flour, annually produced 10-23 times as
much food by weight per acre as wheat, and half that capacity
compared with the potato, after considering an immature period of
up to 15 years and a stem length of up to 20 ft. Contemporary
yield levels place productivity at a level where a palm could
sustain a family from 5-6 weeks in Sarawak, to three months in
Indonesia from a wild palm of 30 m length (Morris 1953; Burhamzah
1970).

Village methods of production during the Occupation, using
the plank-grater and kneading or trampling, which were a considerable improvement on the even more labour intensive methods employed elsewhere in the archipelago, could yield up to a quarter ton
of crude sago per palm in southwest Malaya, and under one-fifth
ton in Sarawak. This flour is not as laundered as commercial
sago-starch; it contains roughage and is inevitably fermented to
a flavoursome state due to storage overnight for food preparation
the next day or two.

While the majority of earlier accounts on sago production
for food assert that sagopalms were not planted, that such a
productive larder for most communities was entirely provisioned
from wild palms is barely credible, though undoubtedly there are
small groups that manage to do so. Nicolaisen (1976) attested
to even the settled Penan's practice of planting hill-sago, a
similarly suckering Eugeissona sp. Formerly, in Tirun Island
"the mangrovy land was planted with sago in great numbers every
year to prevent any deficiency as they are long in growing"
(Dalrymple 1849). Baring-Gould and Bampfylde (1909) observed
in Sarawak "jungles of the cultivated palm where fifty years ago

there were but patchy plantations". The "cultivated productions"
of Minahassa in north Sulawesi included sagopalms (Crawfurd 1856),
and the Marind transferred thornless forest seedlings to their
village palmeries (Barrau 1959). Beyond the Sepik plain "sago
is generally collected ... in the hills also from originally
planted groves ..... it is more or less a part of the subsistence
cultivation system in that the palms are commonly planted in
suitable wet spots, particularly slump floors, largely untended
and left to natural propagation" (Haantjens et al. 1972). Logan
(1949) noted the sagopalm commonly was "probably descended ...
from some Inchi or zaman daulu, who had better notions when he
planted it".

Except for groves that had vegetated into dense forests
when abandoned or underexploited, useful palmeries were more
likely to have been originally planted and then nurtured, if
haphazardly. "Natural sago is exploited where it is accessible
and vigorous enough to come into flower. Probably the sagopalms
growing under forest conditions are usually not vigorous enough
..... Sago production could be considerably increased by the
clearing of forest trees, planting, and proper spacing of good-
quality sago palms" (Haantjens et al. 1968, 1972). "The longer
the sago-forest is under exploitation, the more superfluous young
trees are cleared away, the better producers the remaining trees
will become" (Held 1957).

Serpenti (1965) conceded that "the growing of sago ... might
be regarded as a possible substitute for agriculture in the
narrower sense of the word". The general impression seems to be
that sago-collecting is a form of forest-gathering; it would thus
be comparable to the "wild" oilpalm in the West African aboriginal
economy, when it was the world's sole source of palm oil. Some
nurturing is made all the more necessary by the palm's protracted
immaturity, so that extensive palmeries must be built up to ensure
that sufficient stocks mature whenever needed. The Papuan has
learnt to debud the palm to boost or conserve its starch reserves
prior to harvesting; the hapaxanthic trait may spring from a
terminal axillary bud rather than the meristem itself (Tupamahu
1909; Corner 1966), hence its apparent resilience to the injury.

While its biology dictates that cultures without extensive swamplands cannot use the palm as a primary staple, for others who have abundant sago conventional cultivation becomes less imperative, sustained or sophisticated. "The population ... depends to a large degree on sago for its diet, and hence uses less land per head agriculturally" (Haantjens et al. 1968). "The inhabitants of the coastal beach ridges appear to have fewer gardens per head. Fishing and sago collecting are consequently of greater importance" (Ruxton et al. 1969). This explains why Wallace could provision himself adequately during his voyage in the mid-19th century in the westerly parts of the archipelago with rice and diverse products of cultivation, but eastwards where "sago forms almost the whole subsistence of the inhabitants", he became increasingly reliant on sago and was unable to procure enough dietary supplements.

A USEFUL COMMODITY

As a precolonial mercantile activity in the archipelago, the intraisland sago trade was the preserve of the maritime Malays. Fragmentary references in Western literature indicate the import- ance of the commodity in the region's traditional commerce, and communities had known famine when this food supply was disrupted. There is ceramic evidence of trade connections between western New Guinea and China of the Han period, augmented from the 13th century by trade documentation in Canton, which included the Moluccas (Maluku) as a productive supply area, of spices parti- cularly. Crawfurd (1820) noted that "sago is most abundant in the islands most distinguished for the production of clove and nutmeg and the geographical distribution seems co-extensive with that of these spices".

Sago production had been encouraged in pre-Columbian times by the Chinese, who developed a lucrative junk trade notably with the Sulu sultanate, which spanned north Borneo across to south Philippines, famed for "sago of the best kind ... and in considerable quantities", one of their "staple native exports" (Moor 1837); landang was the only edible product of vegetable origin of any significance, a grain made from swamp sago. Coveted

by pirates, mariners and traders to sustain them during their
voyaging, and fed to the Papuan slave cargo, the sago epic paral-
lels the tapioca's in the Americas, for it too inadvertently
caused productive coasts to be plundered for food and slaves,
driving the remnant populations into the inland hills.

Crawfurd (1820) surmised that the variety of vernacular
names for the sagopalm throughout the archipelago suggested its
antiquity and former ubiquity as a food source, while in modern
times this is apparent in vestigial settings and peripheral areas.
"Considerable quantities are made at the Poggy Islands, lying off
the west coast of Sumatra, where it in fact forms the principal
food of the inhabitants" (Logan 1849), while across many seas
"in the New Hebrides ... native tradition has preserved the
memory of an early era when the starch was extracted and used
as a food", with a similar retention in the folk memory of the
Fijians (Barrau 1958). As an ancient food and drink derived
from a tree, sago came to the notice of such eminent travellers
as Chau Ju Kua in 1200 and Marco Polo in 1298. Early Melaka
subsisted on sago as its first staple, its manufacture there
being recorded a century before the Portuguese conquered it; the
pre-Melaka East Coast cultures of the peninsula were established
along rivers where sago was already used as a staple food
(Tregonning 1964).

The subsistence pattern of insular Southeast Asia may be
generalised very broadly as having crystallised into three
sectors by the turn of this century.

- the wet-rice culture became entrenched in the western islands,
  i.e. the core Malay world which had been strongly imprinted
  with continental influence some two millennia ago, supplemented by tapioca in particular;
- central Indonesia, where seasonal drought predisposed a
  preference for maize, with fragmentary use also of the
  savanna palms, Borassus and Corypha spp.;
- while the eastern sector, hearth of the Metroxylon spp. and
  centre of its domestication, retained the strongest aroid
  and sago traditions, with widespread adoption of New World
  roots outside the sago niches.

Current sago use is confined to Maluku and New Guinea with their pre-Malay population: about a quarter of the population of Irian Jaya and over 100,000 Papuans are dependent on sago as a primary staple, especially evident in some large river basins of New Guinea. In 1687, Dampier (1697) had observed that "the common food at Mindanao is rice or sago", also using the dry method of sago-making employing mortar and pestle, long since defunct;* only a few remote jungle tribes used any starch palm in this century. In Borneo, Banks (1940) distinguished between the immigrant rice eaters of Malay origin and "the old original Kalimantan stock ... still more or less dependent on sago, many are still uncertain about rice planting". But in the postwar era, even their dietary preference followed the inevitable course already well underway elsewhere by the early 19th century, which had led Crawfurd (1856) to conclude: "Sago is never preferred, even where it is more abundant, to rice". "By and large, sago carries a mildly discreditable social significance ..... To eat sago as a necessity - instead of rice as staple food - is shameful enough" (Harrisson 1970).

Nicolaisen (1976) lamented the modern indoctrination towards a rice diet, noting that the Penan sago diet "seemed to be much better than that of the rice farmers" and personally preferred it, a rare modern echo of De Pages' (1793) sentiment on the south Filipino diet: "Agreeable to the example of the Indians, I lived entirely on roots whose saccharine taste is more pleasant than the uniform insipidity of boiled rice ... I was convinced in the end that they are more nutritious to the body, as well as more relishing to the taste".

Earlier observations on human nutrition are worth recalling in view of the way the modern dietary ethos is diverging from its essence. As Wallace simplified it:

> "The staff of life, such as bread, rice, mandioca, maize or sago ... are the daily food of a large proportion of mankind. ..... To maintain his health and beauty he must labour to prepare some farinaceous product capable of being stored and

---

* Brown (1914) postulated that Papuan sago making expertise was derived from Polynesian ancestors who had learnt to extract food from the pith of tree ferns.

accumulated, so as to give him a regular supply of wholesome food. When this is obtained, he may add vegetables, fruits and meat with advantage".

Forrest (1780) esteemed sago as sustenance: "I have often re-flected how well ... circumnavigators might have fared, when passing this way in distress for provisions, had they known where to find the groves of sago".

Sago is a notoriously poor food, even inferior to the root-crops in nutriment. So devoid of food worth can sago be that to the Malays of Sarawak sago paste, "tumpi, as the food of no repute ... barely counts in Islam dietetic ethic. Being thus accounted for, it can be eaten (as a 'no food') during the daylight fasting period of Bulan Puasa ... whereas to eat rice or any other 'real foods' would break the fast for the zealot" (Harrisson 1970). But perhaps the most astonishing negative effects attributed to sago is that given by Brown (1914), whose obsessive distaste for its baleful influence led him to attribute all native indolence, degeneration and sterility in the archipelago to the sagopalm, calling his sago eaters "hand-to-mouth loafers ... the only care they had was how to secure a wife to make their sago for them", even though "a week's work at the pith of sagopalm would give them enough food for a year"; he considered its flour even more villanous to humanity than the wine, saguir. The deleterious effect of sago on its partakers was also noted by Wallace (1898).

Sago is a desirable basic food because most human diets require an adequate carbohydrate intake, otherwise protein is degraded to maintain caloric efficiency. Within their ecological setting, sago eaters commonly exploit the abundant faunal resources of their watery habitat; most nurture semiwild pigs as well. Thus contrary to popular assertion, a sago diet is not a deficient one in its aboriginal context. MacDonald (1956) even thought:

"A menu perpetually of undiluted sago ... does not appear to do the Melanaus any harm. No doubt medical authorities could demonstrate that it weakens their physique ..... On the contrary, the Melanaus present a pleasing bodily physique ... the women have long been famous for their good looks ... unenviable in the old days when slave-trading pirates roamed the Borneo coasts, for it caused Melanau girls to be especial-ly prized as booty ..... a diet composed wholly of sago is the right recipe for aspiring Venuses".

In its prime, Sarawak sago amounted to three-quarters of world export, much of it produced by these "Belles of Borneo" (Mershon 1922; Morris 1953). But Brown (1914) noted that the Papuan sago working women aged well before their time, because of both under-nourishment and toiling for their apparently shiftless menfolk.

FUTURE OF SAGO

The swamp forests of equatorial Southeast Asia and Oceania represent some of the largest remaining forest reserves. Collier (1979) proposed for Indonesia an inventory of the different types of swamps, to assess their potable water, fishing stocks, etc., an exercise which ought to encompass the entire region, before they are irreparably disturbed.

Currently, the strongest awareness in this region of the food potential of sago arboriculture is evinced in Indonesia, where the populations in the eastern islands have been urged not to destroy their sagoforests. Japan has formed the Sago Palm Research Sub-Committee of the Tropical Plant Resources Research Committee (Sato et al. 1979) in its search for raw materials to fuel its vast starch conversion industry, and is the only indus-trialised country to demonstrate a keen awareness of this huge reserve of one of the most coveted raw materials of the modern food industry.

Excepting Irian Jaya, over a quarter of Indonesia's total area of 160 million ha is swampland, of which 10.5 million ha has agricultural potential; a third of Sumatra's area is under swamp (Table 1; Figure 2). The peaty type of swamp is extensive, mainly in Sumatra, 6.3 million ha, and Kalimantan, 9.7 million ha (Soepraptohardjo and Driessen 1976); the Riau swamp alone is 2.5 million acres. The largest in Sarawak is the Rejang delta, "a vast unbroken swamp" some 0.75 million acres in extent (Baring-Gould and Bampfylde 1909). Essai (1961) singled out the swamp, 2,400 sq miles in extent, between the Huntschein Range and the Sepik estuary in Papua New Guinea, as a potential sago-producing centre if the basic infrastructure were to be provided. The Commonwealth Scientific and Industrial Research Organization

Table 1.   Swamps in Indonesia (mostly alluvial and organic)

Islands	Total area ha	Cultivable area ha
Java, Madura, Bali	4.6	3.7
Sumatra	17.9	1.8
Kalimantan	18.7	3.7
Sulawesi	1.8	1.1
Nusa Tenggara, Maluku	0.5	0.2
Irian Jaya	n.a.	n.a.
Indonesia	43.5	10.5

Source: International Bank for Reconstruction and
Development 1974, Indonesia Agricultural
Sector Survey: Annex 1

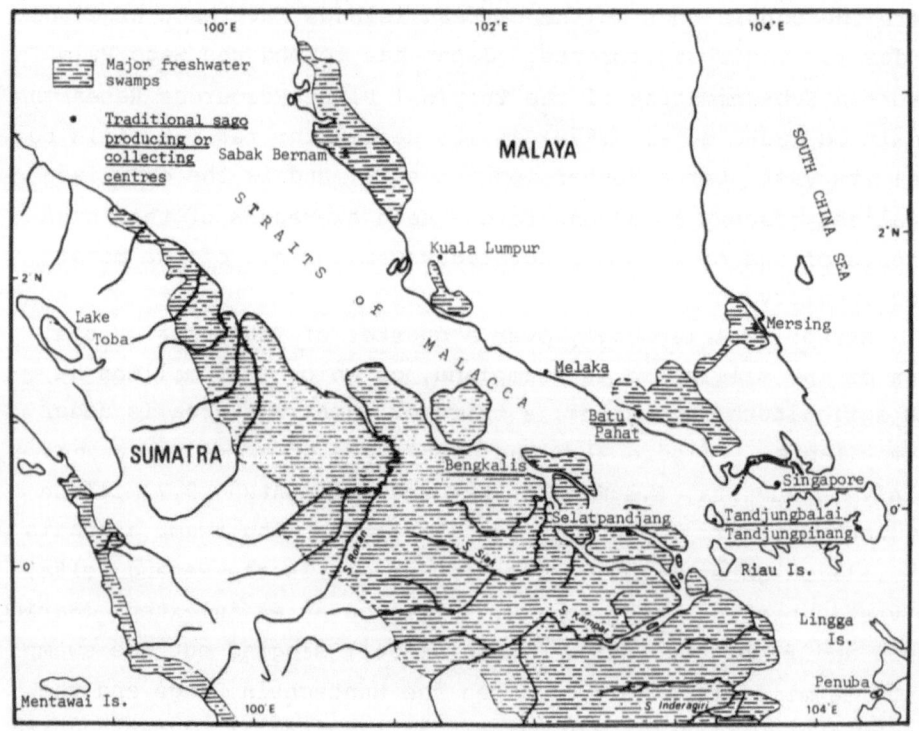

Figure 2.   Major Freshwater Swamps of Malaya and Sumatra

(Australia) land research studies have delimited sago areas of varying degrees of utility to the aboriginal population and for agricultural development (Paijmans 1976).

Logan (1849) had predicted that "the Archipelago can furnish any required amount of meal", but the prevailing view is of "vast sago swamps which offer no opportunities for economic development" (Ryan 1972), valuable only as nature reserves, pastureland, wet-riceland and fisheries (Bleeker 1975); arable cropping is barely possible without elaborate water control. More than a century later this potential remained practically unused, when still it had to be said that "the huge Metroxylon forests of New Guinea represent enormous potential resources in starch for export and industrial uses" (Barrau 1959). Whitmore (1975) believed that the pure sagoforest could have a commercial value not inferior to the lowland rainforest. Haantjens et al. (1968) have pointed out that sagoswamps, where poorly drained, are unsuitable for cultivation; further, "land development is handicapped by the dense vegetation of sago palm ..... Although the sago resources are very limited at present, this land offers excellent growing conditions for this crop if it is planted on cleared land or in thinned-out forest".

In 1923, Winstedt's (1923) assessment is worth recalling: "Crops, whose successful cultivation is bound up with the problem of the production of power alcohol as a substitute for petrol, are the starch-yielding crops". This development counters Brown's (1914) homily that "where sago grows there is and there can be no progress". But it does require a reorientation in crop exploitation strategy: to select suitable crops for particular habitats instead of altering the environment to suit favoured crops, which has been the universal practice so far in land development policies in the intertropical zone. The sagopalm is one of very few crops that could support an ecologically sound system of commercial farming in the equatorial swamplands of Southeast Asia, viz. agroforestry, as has been demonstrated in preindustrial, precolonial production and trading systems for some two millennia.

28

Figure 3. Nipa Fringing a Brackish Tidal Creek; landwards
Rhizophora-Bruguiera mangrove forest colonise tidal
flats; a coconut grove occurs above hightide level.

Figure 4. Mixed Nipa-Mangrove on Brackish Tidal Flat at High
Tide, with gardens of coconut, sugarcane, banana
and sweet potato established on crab mounds formed
abundantly by a tidal channel.

Illustrations courtesy of the Commonwealth
Scientific Industrial Research Organization, Canberra

Appendix

In Dutch studies of the land use of the Indonesian islands underestimation of the nipa-sago forest zones is pointed out by the Koninklijk Nederlandsch (1938): "Extensive bogland is often found in Sumatra and Borneo immediately behind the mangroves, with freshwater bog forests, which are often confused with the real mangrove and are but too often combined with them under the name of 'tidal forests'". The nipa ecotone is common in the transition zone between mangrove and freshwater vegetation types in New Guinea. As the mangrove moves further out to sea, the freshwater swamp forest shifts seawards with decreasing brackish-ness; decreasing swampiness might cause alluvium forest to move too. Harrisson (1970) aptly described this humid zone as "dry land ... in the wettest sense ... The Land Behind the Sea".

In New Guinea, Nypa fruticans covers extensive lowlying parts of estuarine tracts subject to brackish flooding, and also lines tidal creeks fringing mangrove forests (Figure 3); it becomes scarcer with decreasing tidal influence. At the lowest sites, nipa forms dense pure stands, in which the canopy of fronds grows up to 35 ft high, with little undergrowth. On slightly higher ground scattered trees emerge above the nipa, while in some swales and along tidal creeks, it is often found in juxtaposition with stunted sago less than 3 m tall.

The nipapalm "probably does much to stabilise muddy banks of estuaries through its extensive and deep root system" (Paijmans 1976). The vegetation is sensitive to environmental conditions, salinity of the water probably being one of the major influences (Ruxton et al. 1969). In several nipaswamps, a rectilinear network of close spaced tidal channels surround small islands, up to 20 ft in diameter, with central depressions and raised margins breached by a few drains, which are thought to be the result of a combination of tidal movements that deposit sediment, and crab mound-building, which significantly builds and transforms the landscape and vegetation of nipa flats.

The crabs, Thalassina anomala, further aid the accretion process by removing sediment from intermound areas and channel sides, thus making room for further deposition; they also mix

Table 2. Nipa Area in Peninsular Malaysia, 1968 and 1977

State (acres)	1968	1977
Johor	1,870	277
Kedah	2,596	802
Kelantan	868	479
Melaka	310	46
Negri Sembilan	220	0
Pahang	4,582	1,624
Penang/Seberang Prai	2,088	579
Perak	6,405	427
Perlis*	0	162
Selangor	3,050	52
Trengganu	2,916	431
Peninsular Malaysia	24,905	4,879

* During 1969-1973 acreage averaged 500 or more annually.

Source: Min. Agric. Co-op., Malaysia. Statistical
Digest 1968, 1977; Kuala Lumpur

the mud to great depths. The conical mounds may be built up to
2 ft above high tide level which become interconnected to form
small islands, the higher parts of which are colonised later by
a varied freshwater vegetation; the islets or levees decrease
in number away from the channels. In places they are cleared
by the local population for gardening (Figure 4), the only areas
where cultivation of any sort is possible; but the nipapalm and
various mangrove species persist in the channels. In the tidal
flats, crabs may be the regular protein source to the inhabitants,
who often use sago as their staple food where this palm is
accessible.

CSIRO land use surveys revealed that the photo images of
nipa and sago vegetation are similar, but with the nipa a little
lighter in tone, often speckled rather than blotched, and height
tends to be regular. "For the purpose of land use planning it
may be important to know the extent of the brackish zone .....
in the Purari delta, the presence of Nypa was the main criterion
in interpreting a boundary between mangrove and tidal freshwater

swamp forest ..... Increasing frequency of sago palm often helps
in establishing a boundary" (Ruxton et al. 1969). "It is a
feature of halophytic vegetation, which is almost invariably
floristically stable, that it has arisen in most cases as an
adaption of some insufficiency" (van den Berg 1973).

As Paijmans (1976) pointed out, this monotypic palm is "a
potential, relatively cheap source of sugar and alcohol". The
nipa area in Peninsular Malaysia has shrunk considerably in the
last decade (Table 2) due to drainage mainly for rice irrigation
schemes, and port and land development. This delicately balanced
habitat, once destroyed, cannot be fully rehabilitated, and many
of its potential resources would become lost. Contemporary
development philosophy appears to work to this end, for no ASEAN
country has a systematic wetland development strategy which would
indicate its recognition that the equatorial zone shelters
extensive belts of swampland that defy conventional exploitation.
The nipaswamp, like the sagoswamp, could be preserved as a
reservoir of alternative energy.

REFERENCES

AIKMAN, R.G. 1959. Melanaus, op. cit., ed. HARRISSON.

BANKS, E. 1940. The natives of Sarawak, J. Malay. Br.
     roy. Asiat. Soc. 18(2).

BARING-GOULD, S., and C.A. Bampfylde 1909. A History of Sarawak
     under its Two White Rajahs, 1839-1908; Sotheran, London.

BARRAU, J. 1959. The sago palms and other food plants of marsh
     dwellers in the South Pacific islands, Econ. Bot. 13(2).

BLAKE, D.H., et al. 1973. Land-form Types and Vegetation of
     Eastern Papua; Commonw. Scient. Indus. Res. Org. Land
     Res. Ser. no. 32, Canberra.

BLEEKER, P. 1975. Explanatory Notes to the Land Limitation
     and Agricultural Land Use Potential Map of Papua New
     Guinea; Commonw. Scient. Indus. Res. Org. Land Res. Ser.
     no. 36, Canberra.

BREWSHER, R.A. 1959. The Bisaya group, op. cit., ed. HARRISSON.

BROWN, J.M. 1914. The Dutch East. Sketches and pictures;
     Kegan Paul, London.

BURHAMZAH 1970. An economic survey of Maluku, Bull. Indon.
     econ. Stud. 6(2).

32

BURKILL, I.H. 1935. A Dictionary of the Economic Products of the Malay Peninsula; (1966 rep.) Min. Agric. Co-op., Kuala Lumpur.

CAMERON, J. 1865. Our Tropical Possessions in Malayan India; Smith Elder, London.

CANTLEY, N. 1886. Notes on economic plants (Appendix C), J. Str. Br. roy. Asiat. Soc. 18.

COLLIER, W.L. 1979. Social and economic aspects of tidal swamp land development, Symp. Tidal Swamp Development Aspects 1979, Palembang (mimeo.).

CORNER, E.J.H. 1966. The Natural History of Palms; Weidenfeld and Nicolson, London.

CRAWFURD, J. 1820. History of the Indian Archipelago; Constable, London.

CRAWFURD, J. 1856. A Descriptive Dictionary of the Indian Islands and Adjacent Countries; (1971 rep.) Oxford Univ. Press, Singapore.

DALRYMPLE, A. 1849. Essay towards an account of Sulu, J. Indian Arch. east. Asia 3.

DAMPIER, W. 1697, vol. 1, op. cit., pub. FILIPIANA 1971.

DE PAGES, P. Vicomte 1793., op. cit., pub. FILIPIANA 1971.

ELMBERG, J.E. 1968. Balance and Circulation. Aspects of tradition and change among the Mejprat of Irian Barat; Stockholm Ethnograph. Mus. Mon., pub. no. 12.

ESSAI, B. 1961. Papua and New Guinea. A contemporary survey; Oxford Univ. Press, Melbourne.

FILIPIANA Book Guild 1971. Travel Accounts of the Islands (1513-1787); Manila.

FORREST, T. 1780. A Voyage to New Guinea and the Moluccas from Balambang; (1969 rep.) Oxford Univ. Press, Kuala Lumpur.

FOX, J.J. 1977. Harvest of the Palm. Ecological change in eastern Indonesia; Harvard Univ. Press, Cambridge.

HAANTJENS, H.A. et al. 1968. Lands of the Wewak-Lower Sepik Area, Papua New Guinea; Commonw. Scient. Indus. Res. Org., Land Res. Ser. no. 22; Canberra.

HAANTJENS, H.A. et al. 1972. Lands of the Aitape-Ambunti Area, Papua New Guinea; Commonw. Scient. Indus. Res. Org., Land Res. Ser. no. 30, Canberra.

HARRISSON, T. (ed.) 1959. The Peoples of Sarawak; Sarawak Mus., Kuching.

HARRISSON, T. 1970. The Malays of South-west Sarawak before Malaysia. A socio-ecological survey; Macmillan, London.

HELD, G.J. 1957. The Papuas of Waropen; Martinus Nijhoff, The Hague.

The INDONESIAN TIMES, 15 Sep. 1979: "Sago said a possible substitute for oil".

KONINKLIJK NEDERLANDSCH AADRIJKSKUNDIG GENOOTSCHAP 1938. Atlas van Tropisch Nederland; Batavia.

LÖFFLER, E. et al. 1972. Land Resources of the Vanimo Area, Papua New Guinea; Commonw. Scient. Indus. Res. Org. Land Res. Ser. no. 31, Canberra.

LOGAN, J.R. (ed.) 1849. Sago, J. Indian Arch. east. Asia 3.

MABBUTT, J.A. et al. 1965. Lands of the Port Moresby-Kairuku Area, Territory of Papua New Guinea; Commonw. Scient. Indus. Res. Org. Land Res. Ser. no. 14, Melbourne.

MACDONALD, M. 1956. Borneo People; Jonathan Cape, London.

MAHER, R.F. 1961. New Men of Papua. A study of culture change; Univ. Wisconsin Press, Madison.

MERSHON, N. 1922. With the Wild Men of Borneo; Pacific Press Pub. Assoc., Mountain View (California).

MORRIS, H.S. 1953. Report on a Melanau Sago Producing Community in Sarawak; Col. Res. Stud. no. 9, Col. Off., London (mimeo.).

MOOR, J.H. 1837. Notices of the Indian Archipelago and Adjacent Countries; (1968 rep.) Frank Cass, London.

NICOLAISEN, J. 1976. The Penan of the Seventh Division of Sarawak; Past, present and future, Sarawak Mus. J. 24(115).

PAIJMANS, K. (ed.) 1976. New Guinea Vegetation; Austral. Nation. Univ. Press, Canberra.

RUXTON, B.P., et al. 1969. Lands of the Kerema-Vailala Area, Territory of Papua and New Guinea; Commonw. Scient. Indus. Res. Org. Land Res. Ser. no. 23, Melbourne.

RYAN, P. (gen. ed.) 1972. Encyclopaedia of Papua and New Guinea (2 vol.); Melbourne Univ. Press/Univ. Papua and New Guinea.

SATO, T., et al. 1979. Special report, Japan. J. trop. Agric. 23(3) (all papers exclusively on sago palms).

SEEMAN, B. 1856. A Popular History of the Palms and their Allies; Lovell Reeve, London.

SERPENTI, L.M. 1965. Cultivators in the Swamps. Social structure and horticulture in a New Guinea society; van Gorcum, Assen.

SOEPRAPTOHARDJO, N., and P.M. Driessen 1976. The lowland peats of Indonesia. A challenge for the future, Proc. Sem. Peat and Podzolic Soils and the Potential for Agriculture in Indonesia, Tugu; Soils Res. Inst. bull. 3, Bogor.

STANLEY, B., W.H. Allsop and F.B. Davy 1978. Fish Farming; International Development Research Centre, Ottawa.

TAN, K. 1979. The palm economy in particular the swamp sago economy in Southeast Asia, J. d'Agric. trad. Bot. appl. 26(2).

THOMSON, J.P. 1892. British New Guinea; George Philip, London.

TREGONNING, K.G. 1964. A History of Malaya; Eastern Univ. Press, London.

TUPAMAHU, J. 1909. Sagoe en sagoepalmen, Bull. Koloniaal Mus. te Haarlem 44 (illustrations).

VAN DEN BERG, C. (ed.) 1973. Irrigation, Drainage and Salinity. An international source book; Hutchinson/FAO/ UNESCO, London.

WALLACE, A.R. 1898. The Malay Archipelago; (10th ed.; first pub. 1869) Macmillan, London.

WHITMORE, T.C. 1975. Tropical Rain Forests of the Far East; Clarendon, London.

WINSTEDT, R. (ed.) 1923. Malaya, The Straits Settlements and the Federated Malay States; Constable, London.

POTENCY OF SAGO AS A FOOD-ENERGY SOURCE IN INDONESIA

R. SOERJONO

DISTRIBUTION OF SAGO

Many palms produce pati (amylum/starch), e.g. aren, Arenga pinnata, in Java, also found in Sumatra and Kalimantan; in the last few months it has been found that Eugeissona sp. also produces amylum. But the pure amylum comes only from the Metroxylon spp. which the Minangkabau call rumbia, the west Javanese kira, central and east Javanese bulung or kresula, and the Amboinese lapia or napia.

Up till the present there is no reliable data on the extent of the sagoforests in Indonesia; an inventory in this field has recently been started. It is estimated that the sagoforests in Maluku extend to about 500 thousand hectares with a production capacity of about 800 thousand tonnes of flour or an annual yield equivalent of about 1.6 t/ha. In Irian Jaya, the sago area covers about 7.5 million ha; applying the Maluku estimated yield, this area has an annual production potency of 12.3 million tonnes of carbohydrates.

TYPES OF SAGO PALMS

The sago species that are twice-fruited are M. filare and M. elatum, which grow at rather higher places and produce only a little sago. Five hapaxanthic Metroxylon palms are found in Indonesia, of which only M. sagu and M. rumphii have economic value as starch producers.

M. rumphii Mart., called sagutuni or lapia tuni in Ambon, reaches 10 m in height. Its leaflets are light green with a long leafstalk, with spines 1-4 cm long at the lower end, arranged diagonally. It has a shallow rooting system and suckers freely from the base. Its girth is 50-60 cm, with bark 2-3 cm thick; its soft marrow produces a delicious flour, yielding about 500 kg/palm.

M. sagu Rott., is called lapia mulat or sagu perempuan (female sago) in Ambon; the latter name is given probably because this species has no spines. It is also abundant in Riau, Sumatra. The trunk has a soft marrow, and the leaves are long and pointed at the end. It yields a fine tasty flour much liked by the Amboinese; yield is about 200 kg/palm. Flour from this palm is much exported.

M. longispinum Mart., found in the Moluccas, is called "red sago". It is less desired for food because it produces a small amount of sago, 200 kg/palm. The palm is identified by the small trunk, long spines at the leafstalks, with rather small leaves that are straight and stiff.

M. microcanthum Mart., named "rattan sago" in the Moluccas, grows abundantly in the Ceram region, while in Ambon it is less favoured; it is identified by spines that are shorter than those of other spiny species. Its special feature is that it has marrow which will not become acidulated during harvesting in the forest.

M. sylvestre Mart., has a red and hard marrow with an indelicate taste. Leaves are dark green, crooked at the end. The Ceramese do not plant this species, but in Halmahera it is found in great numbers where it is harvested because it produces a long trunk and also reddish sago.

SAGO PRODUCTION

Since sago has been known from early times and is planted in Irian Jaya, Maluku and Riau, the Indonesians already knew the principles of sustained yield. The palm can be propagated from seed or root sucker; the latter practice gives the best results, as was done in Irian Jaya and Riau.

The soils should be tilled before planting, since the sucker should be planted directly after it is cut from the mother-palm; planting distance is 5-6 m. The sucker should be firmly set but not too deep into the soil, because in a flooded planting it is liable to die if submerged too long. On fertile soil, it reaches maturity 9-15 years after planting.

Maturity in the sagopalm is identified by the changing colour of the leafstalks and the emergence of the flower spike. This is the optimum time to fell the palms down because the starch content reaches the highest amount; to sample the starch content, the people in the eastern part of Indonesia usually drill into the trunk.

The domestic method of sago production in Maluku has been described elsewhere. The starch is dried before it is stored or quickly taken to the market for sale; if it is to be kept for a long period, the starch must be washed, sieved and dried again to remove the acid smell and to obtain a whiter substance.

The factories of Riau and the surrounding area make sago-starch into pellet, grain, pearl or flake for the international market. Pearl sago is formed by shaking the moist starch on an iron plate, then dried on a heated metal plate and polished with tengkawang oil; the size of the grain can be varied with the aid of sieves of different meshes. Flake sago is made by squirting wet sago onto a very hot copper plate.

There are manufactories that produce glucose, alcohol, and dextrine from sago; in small industries sago is used in tapioca mixtures.

PROMOTING SAGO VALUE

Through appropriate silvicultural treatment and prompt pro-cessing methods, the potency of the sagoforests in Indonesia could be multiplied. Though sago played a great part in fulfilling the food needs of the country before World War Two, its use has since declined, largely due to the lack of modernisation in production methods. To raise sago consumption, technical improve-ments must be made, either in the production of the flour or the processing of sago foodstuffs.

If one hectare of land could carry 625 palms, i.e. 62,500 palms/km^2, then 22,500 km^2 of land (150 x 150 km) would contain 1,406,250,000 palms. Harvesting at the age of 12 years would allow 117,187,500 palms to be felled annually; with each palm

producing 350 kg sago, the annual production would be 41,015,625
tonnes. This amount could fulfill the carbohydrate needs of
41 million persons, assuming an annual consumption of 1000 kg
per caput. Therefore, bearing in mind the food crisis in
Indonesia, it is recommended that sago be accorded a more extensive role as a primary food staple.

According to Dr. Masri Singarumbun, the annual production
of raw sago in Irian Jaya could even reach 77 million tonnes,
viz. the potency of sago in Irian Jaya is 15.4 times greater
than Maluku. According to the Director of the Demography Department of Gadjah Mada University, if the potential sago production
were to be used for food, dependence on imported rice and maize
could be considerably reduced within a short time.

Sago production currently is still well below optimum capacity, especially considering that it employs inefficient extraction methods. The volume can be vastly increased by applying
modern techniques of production that would also stimulate greater
consumption of its products. The National Census indicated
that some Indonesians still derive 30-40% of their carbohydrate
intake from sago; formerly the carbohydrate consumption reached
60%.

NUTRITIVE VALUE OF SAGO

Sago has a high carbohydrate content, ca. 95% (dry), compared
with red rice 75% and maize 64%, but it has a much lower nutritive
value; it is also deficient in vitamins, especially A, B and C.
Thus if taken as a staple food, sago will give for example only
3.2 g albumen, cf. rice 40 g and maize 50 g; but the calorie
value of these foods is the same, at about 360 Cal per 100 g
dried matter.

Although sago is regarded as a poor food, it has been found
that people who depend on sago, in Irian Jaya, Maluku and neighbouring islands, have a firmer body than those whose main food
is rice or maize. This is because sago eaters add fish and
vegetables to their diet. With suitable supplementary foods
then, the sago diet can have the same quality as cereal diets.

SAGO AND THE FOOD-ENERGY SHORTAGE IN INDONESIA

SOESARSONO WIJANDI

AWARENESS OF SAGO AS A FOOD STAPLE

Sago or sagu in west Java and north Sulawesi is the starch extracted either from Arenga pinnata or Metroxylon spp.; for most people in Indonesia however, it means the processed food called sagu ambon or sagu lempeng from Maluku and Irian Jaya.

Most of the information on sago originates from Maluku, particularly Ambon, as other sources are scarce due to limited investigations on the subject. The present awareness on the possible role of sago as a major food source, and growing literature on sago presented in newspapers and seminars, are stimulating research on and development of sago use in Indonesia. There is a national project on sago use.

Sago is useful in different ways. In the western parts of Indonesia starch is a common commodity produced from the sagopalm; in west Java sago from Arenga and Metroxylon spp. is sold usually for making noodles and cendol. But mainly the palm is used for thatching purposes rather than for food, because there the need for shelter and an income exceeds the demand for the starch. In south Kalimantan ducks are commonly fed on the shredded pith of the trunk.

In eastern Indonesia, sago is one of the staple foods of the local people; it is the main food staple for 30% of the population in Maluku and 20% in Irian Jaya (Tables 1 and 2). It is estimated that more than 100 thousand tonnes of processed sago, mainly sagu lempeng produced in Maluku, are annually sold to and consumed by people from other islands.

PRODUCTION OF SAGO AND FOOD DERIVATIVES

The sagopalm is mostly felled by men. The trunk is cut

Table 1.   Sago Consumption in Maluku

District	Population thousand	Sago Consumption %	
		Main diet	Secondary diet
North Maluku	360	35	46
Halmahera	102	59	32
Central Maluku	500	78	11
Ambon	120	4	22
Southeast Maluku	172	15	23

Source: Office of the Department of Industry, Ambon 1978

Table 2.   Staple Carbohydrate Food in Irian Jaya

Food Source	Proportion %	Locality
Sweet potato	30	central upland
Xanthosoma sp.	30	low upland
Sago	20	swamp
Rice	20	town

into certain lengths and split in half; the inner part is chopped and shredded with nani, a tool made of hard wood or bamboo, tipped with metal or stone.   This work is done mostly by women in Irian Jaya but strictly by men in Ambon.

In general, there is no significant difference in the methods of extracting sago from one area to another in Indonesia already described elsewhere.   The whole task from felling the palm to transporting the basket sago needs no less than 112 manhours per palm.   Usually a good palm can supply enough food for a family for up to three months.

Wet sago is packed in baskets of sago leaves, tumang, each weighing about 12 kg; the larger one weighs about 30 kg.   The basket sago is kept wet all the time to keep the starch fresh, for making papeda, and can be stored for six months or longer by occasional watering; mostly it is stored for no more than two months.

Loss of starch during transportation and storage of tumang can be as high as 25%.   The sago left in the stump and top of

the felled trunk may reach 20%; further loss occurs if the trunk
is allowed to deteriorate before the shredding process is
completed.

There are several kinds of processed food made of sago.
Research on the further development of sagu lempeng and buburnee
is needed, because they are popular staple foodstuffs in several
parts of eastern Indonesia. Buburnee is a rice-like sago food
that resembles gari in Ghana.

Sagu lempeng is one of the most interesting foods made of
sago; it has a relatively high resistance to common damaging
factors during transportation and storage, because so far there
has not been any report of spoilage of sagu lempeng caused by
moulds and insects. It is relatively non-hygroscopic but ready
to swell quickly when it is dipped in liquid or beverage that
usually goes with it in the diet; fishermen in particular prefer
it to other foods.

Papeda can only be made from fresh sago and may be considered as the nasi (cooked rice) of the local people. North
Sulawesi produces the wellknown bagea menado, sticks of sago
wrapped in sago leaves and then smoked; bagea is a common name
for processed food in many forms that resemble cookies.

SAGO FOOD POLICY

The food shortage is still the main problem in Indonesia,
for she imports annually about two million tonnes of rice and
about one million tonnes of wheat, which appropriate substantial
resources that she can ill afford. Unfortunately, the archipelago is also frequently beset by many kinds of natural disasters
such as earthquake, flood and drought; the government provides
not less than 100 thousand tonnes of food, mostly rice, for
such emergencies every year. In many cases, giving rice as
emergency food in some areas raises other problems, especially
where the population is not rice-oriented.

On the other hand, sago is abundant and largely untapped
yet. To save capital resources and to avoid social problems,
for such a situations it is important to develop emergency foods
based on sago. This kind of food should at least meet the

qualifications of storability, palatability, nutrition and price. Sagu lempeng meets these qualifications except the nutritional value; research on this aspect is undertaken now in Indonesia.

Research and development to improve traditional methods of sago processing are essential, and have just begun in Indonesia. Internal and international cooperation among educational and research institutions is needed, if the national programme on sago use should be implemented.

The traditional methods of sago production are tedious and most young people do not like the work. More efficient equipment and methods should be developed to make the task more attractive to them, although for the time being large scale and fully mechanised exploitation is not favoured for social reasons. Collecting, transportation and storage systems should also be studied if the government stockpiles sago for food use.

REFERENCES

AMBON, Dep. Industry 1978. Riset Sagu.

SOESARSONO WIJANDI et al. 1979. Report on Sago Survey in Irian Jaya and Maluku.

BAS YOUWE 1979. Masalah Pangan terutama Sagu di IRJA.

THE COMPARATIVE NUTRITIONAL ROLES OF SAGO AND CASSAVA IN INDONESIA

LIE Goan-Hong

USE OF SAGO AND CASSAVA AS STAPLE FOODS

If the sagopalm as a food source must already have been known in Indonesia since prehistoric times (Avé 1977), the cassava plant however has become known only recently. Introduced into Indonesia at the beginning of the 19th century from South America, cassava had already become a very important foodcrop at the turn of this century, ubiquituously cultivated and consumed throughout the archipelago. This was made possible after the Agricultural Station in Bogor in 1918 had succeeded in selecting the proper varieties, using seedlings (Heyne 1950).

The sagopalm has not become a foodcrop of similar geographical extent most probably not only because of its restrictive habitat, but also its physiological characteristics, mode of exploitation, among other restraints discussed at the First Symposium. Table 1 shows the variety of food derivatives from both plants.

A picture of the total production and consumption of sago ambon and cassava in Indonesia in 1976 can be seen from Table 2. That no data can be found regarding the production of sago ambon, not even in the statistics on minor forest products, must be due to the fact that it is not an export commodity and hence of local importance only. Actually, sago ambon is used as food in geographically limited regions by ethnic groups with more or less subsistence economies; it has still no significant role in inter-insular trade.

The total quantity of both foods consumed can be reliably estimated by multiplying the average daily per caput consumption with the 1976 population of each island (Table 3). Assuming

W.R. Stanton and M. Flach (eds.), SAGO. The Equatorial Swamp as a Natural
Resource. Proceedings of the Second International Sago Symposium. All rights reserved.
Copyright ©1980 Martinus Nijhoff Publishers, The Hague/Boston/London.

Table 1.   Traditional Foods from Sago and Cassava

Sago (Metroxylon spp. except where stated)

"Cabbage", palmiet (in Dutch), inner shoot of crown:
    fruit or snack or substitute for bamboo shoot

Sap, tapped from male inflorescence of Arenga pinnata:
    fresh - saguwir or tuak
    boiled, concentrated - palm sugar
    fermented - vinegar; distilled spirit, sopi or cap tikus

Fruit (A. pinnata):
    inner kernel, cooked fruit dessert in syrup, buah atap
      or kolang-kaling

Trunk:
    flour, direct use - cooked into porridge, papeda; baked
      into sagu buluh (in bamboo), sagu ega (in leaves),
      sagu senole (with grated coconut)

    long-term use - roasted as briquettes in earthen moulds
      sagu-bi or sagu lempeng, cookies, bagea manis/asin,
      sweet/salty (wrapped in leaves), bagea ternate (with
      milled canarium nuts)

Others:
    larva or grub of sago beetle, Rhyncophorus bilineatus -
      eaten raw, roasted or fired
    jamur sagu (mushroom grows on ampas, pith residue)

Cassava (Manihot esculenta)

Leaf:
    fresh, young - cooked or boiled as vegetable

Root:
    peeled, fresh - boiled, fried, roasted and eaten as such
    rasped - mixed with grated coconut and palm sugar and
      steamed, as a snack
    boiled, fermented - tapé (central Java) or peuyeum (west
      Java), snack food, consumed fresh, fried or baked

    peeled, dried - gaplek
    pounded and steamed - tiwul, used as staple food; also
      mixed with rice or maize grits
    flour - used as snack and in drinks; cakes, cendol
    refuse, ampas or onggok - used like breadcrumbs in frying
      of tempeh; steamed, added to oncom
    outer peel - steamed, fermented for 3-4 days into dagé,
      a sidedish; used as emergency food

that sago ambon is produced for local use only, the total consumption must be the same as total output.

    The total production and consumption of cassava is very high (Table 2), whereas that of sago ambon is small and limited to the eastern part of Indonesia, viz. Sulawesi, Nusa Tenggara

Table 2. Estimated Production and Consumption of Sago
Ambon and Cassava in Indonesia, 1976

| Region | Production[a] | | Human Consumption[b] | |
| | Sago ambon (dry) | Cassava (peeled dry) | Sago ambon (dry) | Cassava (peeled dry) |
		thousand tonne		
Indonesia[c]	not available	3,717.0	97.9	1,448.0
Java-Madura		2,734.2	0.6	828.5
Outer Islands[c]		982.7	99.3	621.4
Sumatra		409.9	7.6	403.6
Kalimantan		81.6	0.6	47.5
Sulawesi		207.3	66.0	93.5
Bali-NTB-NTT-Maluku		283.9	25.8	81.6
Irian Jaya		7.8	not available	

Note: Discrepancies in sum totals are due to rounding off of
figures. Figures presented are calculated from the
following sources:

[a] Production: Central Bureau of Statistics 1978: 369.

Conversion of factors used for cassava root (ref. Table 6):
fresh root: refuse 27%, moisture content 65.5%,
dried peeled: moisture content 15.7%.

[b] Consumption: based on data given in Tables 3 and 4.

[c] Irian Jaya not included
NTB = Nusa Tenggara Barat
NTT = Nusa Tenggara Timur

and Maluku. The high cassava production and consumption are
spread across Indonesia, but Java by far is the largest producer
with 74%. Of the total production of Java, 30% or 829 thousand
tonnes is directly consumed by the population, while the remaining
70% must be used for other purposes, such as for export. On the
other hand, Sumatra which produces 11% of the total cassava, uses
it all for human consumption; Sumatra has indeed the highest
cassava consumption viz. 152 g/person daily, compared with an
average of 74-90 g in all other islands, including Java (Table 3).

The total consumption of sago ambon is the highest in
Sulawesi with 66 thousand tonnes, as compared to 26 thousand
tonnes for Bali-Nusa Tenggara-Maluku. This finding seems to

Table 3. Average Daily per Caput Consumption of Staple Foods in Indonesia, 1976

Region	Energy (A) + (B) g	Protein (A) + (B) g	Cereals (A)			Non-Cereals (B)			
			Rice g	Maize g	Wheat flour g	Cassava fresh root E.P. g	Sweet potato E.P. g	Yam E.P. g	Sago ambon g
Indonesia*	1353+181	26.9+1.7	322.14	51.55	2.86	99.00	19.78	5.46	2.00
Java-Madura	1238+150	24.7+1.4	293.43	49.68	1.43	89.08	12.53	5.10	0
Outer Islands*	1562+236	31.0+2.2	374.00	55.17	5.43	116.67	32.93	6.19	5.57
Islands									
Java-Madura	1238+150	24.7+1.4	293.43	49.68	1.43	89.08	12.53	5.10	0
Sumatra	1440+283	27.3+2.8	394.29	2.35	3.29	152.42	45.58	1.58	0.86
Kalimantan	1611+134	30.6+1.3	437.57	1.68	8.14	73.57	18.80	2.79	0.29
Sulawesi	1680+236	35.0+2.1	336.71	122.14	9.43	87.37	18.18	22.10	18.43
Bali-NTB-NTT-Maluku	1716+187	36.4+1.7	318.71	154.63	5.29	84.78	25.19	3.89	8.0

Note: Figures given are not rounded off to enable further statistical calculations as needed

* Irian Jaya not included, for which information is unavailable

E.P. = edible portion

Source: Based on data from Central Bureau of Statistics, 1977

contradict the statements made by various authors (Postmus and van Veen 1949; Avé 1977; Ellen 1977; Sastrapradja and Mogea 1977) who reported that Maluku, especially the island of Seram, is the centre of sago ambon consumption. As the present staple foods of Bali, West and East Nusa Tenggara are rice and maize, a changeover that has been described by Avé (1977), one may assume that the volume of 26 thousand tonnes refers to the province of Maluku only. Taking the population of Sulawesi and Maluku into consideration (Table 4), the average annual per caput consumption of sago ambon in 1976 in Sulawesi would be only 6.7 kg and in Maluku 20.7 kg; this would agree more with the statements made by those authors.

According to the Under-Minister for Food, not yet fully used are the 75 thousand hectares of potential sagoforests dispersed over the northern and central part of the province of Maluku; with an average annual yield of 5 t/ha of sago-flour, Maluku would be able to produce about 375 thousand tonnes of sago-flour annually. The local sago consumption is estimated to be 144 thousand tonnes per year; the balance of 231 thousand tonnes is not exploited. This estimated surplus would be much higher if one applied the yield estimate given by Flach (1977), which is 7-11 tonnes of water-free starch per ha per year under

Table 4. Estimated Population of Indonesia 1975-76

Region	1975	1976
	thousand at end of year	
Sumatra	23,599	24,282
Jawa-Bali	85,903	87,718
Nusa Tenggara Barat	2,461	2,522
Nusa Tenggara Timur	2,565	2,628
Kalimantan	5,773	5,925
Sulawesi	9,562	9,813
Maluku	1,217	1,247
Irian Jaya	1,033	1,058
Indonesia	132,110	135,190

Source: Central Bureau of Statistics 1978: 102-3

wild growing conditions.

The Under-Minister also expressed his concern over the changing trend from sago to rice as the staple food. The community is following the food habits of the civil servants and newcomers, and rice is imported into the region, sold at a relatively low price that makes it easily available (Kompas no. 64, September 3, 1979, Jakarta).

Consumption of Palm Sugar

Another product of the starch palm is sugar, whose energy value is almost the same as that of sago-flour; but palm sugar cannot be used as a staple food. It is worth noting that in the Outer Islands the total consumption of palm sugar (Table 5) is the same as sago ambon (Table 2), all produced locally, i.e. the economic significance of palm sugar is not less than that of palm starch, a fact that is generally not fully realised.

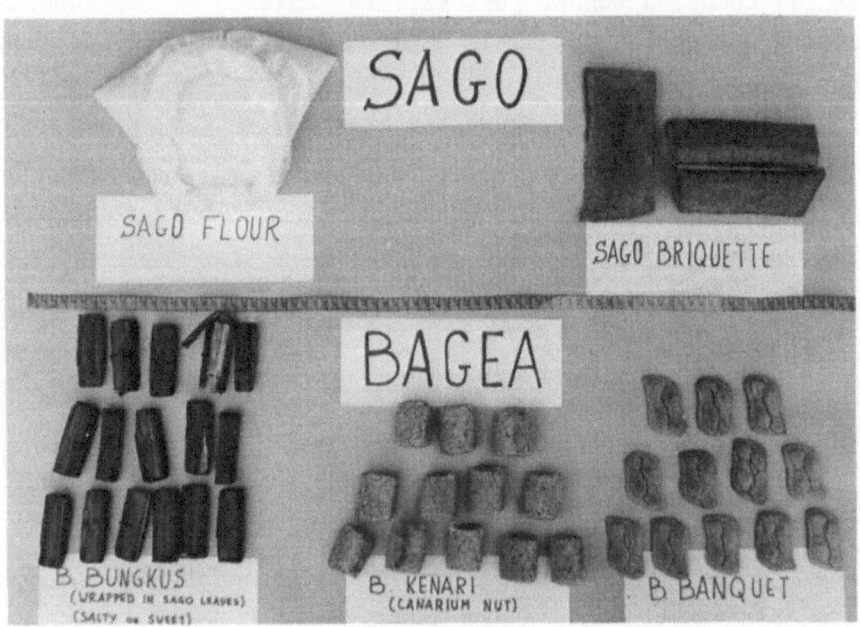

Some Traditional Foods in Indonesia

Table 5. Estimated Consumption of Palm Sugar and Cane
Sugar in Indonesia, 1976

| Region | Total | | Daily | |
| | Palm | Cane | Palm | Cane |
	thousand tonne		g/person	
Indonesia*	334.4	797.1	6.83	16.28
Java-Madura	233.2	399.1	7.49	12.82
Outer Islands*	100.6	401.7	5.64	22.53
Java-Madura	233.2	399.1	7.49	12.82
Sumatra	32.5	190.5	3.67	21.49
Kalimantan	12.0	81.2	5.54	37.53
Sulawesi	39.5	79.4	11.03	22.16
Bali-NTB-NTT-Maluku	16.5	50.0	5.13	15.51

Discrepancies in sum totals are due to rounding off of figures.

* No data available for Irian Jaya

Calculations are based on data as given in Table 4 and Central
Bureau of Statistics 1977.

It has a brown colour and a specific taste, very much liked
by the general population. It is prepared from the sap collected
from the male inflorescence, and concentrated through evaporation
by boiling down in an open iron vessel on an open fire. Palm
sugar production is done at the village level as a home industry
using simple tools, and many villagers earn their living from
this occupation. In Java-Madura gula kelapa or gula Jawa comes
from the coconutpalm, Cocos nucifera; in the Outer Islands, gula
aren is obtained from the aren, Arenga pinnata.

NUTRITIONAL CONSIDERATION IN CHOICE OF STAPLE

The protein content of sago and cassava flour is very low,
about 1% only (Table 6). When eaten as a regular daily food,
a substantial amount of protein from supplementary foods will
be required to meet the total quantity of protein needed by the
body for the maintenance of good health. Under the prevailing
dietary pattern of the general population in Indonesia, sago
and cassava will be unsuitable as a primary staple food, because

Table 6. Approximate Components of Some Starch-Rich Foodstuffs

Food 100 g	Food Energy Cal	Moisture g	Protein g	Fat g	Carbohydrate (incl. fibre) g	Fibre g	Ash g
Cassava, bitter, common (Manihot esculenta)							
Raw: E.P. (refuse 27% of A.P.)	135	65.5	1.0	0.2	32.4	1.0	0.9
Fermented: E.P.	174	56.1	0.5	0.1	42.5	..	0.8
Dried: E.P. (gaplek)	333	15.7	1.4	0.5	80.6	1.2	1.8
Meal	363	9.1	1.1	0.5	88.2	2.2	1.1
Starch	354	12.0	0.5	0.3	86.9	..	0.3
Grits	362	9.2	1.8	0.4	87.9	2.7	0.7
Tapioca pearl	358	9.6	1.4	0.4	87.1	..	1.5
Chips	349	12.0	0.5	0.2	85.9	..	1.4
Sagopalm (Metroxylon spp)							
Meal	357	13.1	1.4	0.2	85.9	0.2	0.4
Sweet potato, yellow (Ipomoea batatas)							
Raw: E.P. (refuse 13% of A.P.)	115	70.7	1.2	0.3	27.1	0.8	0.7
Boiled: E.P.	114	70.7	1.0	0.1	27.4	0.6	0.8
Flour	339	13.2	2.2	0.9	80.8	3.0	2.9
Starch	336	16.5	0.1	0.1	83.2	0	0.1
Steamed, dried	301	23.5	1.8	0.8	71.8	2.2	2.1
Potato, white (Solanum tuberosum)							
Raw: E.P. (refuse 13% of A.P.)	82	78.3	2.0	0.1	18.7	0.4	0.9
Flakes, mashed and dried	361	7.5	6.0	0.6	84.2	1.6	1.7
Starch	332	17.5	0.1	0.1	82.1	0	0.2

A.P. = as purchased
E.P. = edible portion

Source: FAO/US Dep. of HEW 1972

of the relative unavailability (expensiveness) of protein-rich
foods. In general, it can be said that the current staple foods
in Indonesia (Table 3) are rice in the western part and rice-
maize in the eastern, with cassava and sweet potato as additional
staples (Lie et al. 1978).

Rice and maize, having a protein content of about 7% and 9% respectively, are the main providers of not only energy, about 70% of total energy intake, but also protein as well, about 61-73% (Lie 1976). The accompanying foods consist of dried, salted small fish and fermented legume products, such as tempeh from soybean and oncom from peanut presscake; being highly nutritious, these foods will provide the missing protein. Unfortunately however, in many instances the quantity consumed is small as it is relatively expensive. Therefore, under the condition as described above it will be obvious why sago and cassava are not advocated for use as staple foods to replace rice or maize in the diet of the population in general.

It is expected that these starchy foods will have a dele-terious effect on the nutritional status of the population. The description of the classical Gunung Kidul cassava diet and the Seram sago diet in the following part of this paper will be illustrative. Yet sago and cassava are cheap sources of energy for the population, a very important aspect which cannot be ignored. If the total protein in the daily diet is already sufficiently met, then the remaining energy requirement need not necessarily have to come from the relatively expensive rice or maize, but may be obtained more cheaply from cassava or sago. If in general the diet of the Indonesian population is described as a "rice" and a "rice-maize" diet, the two other differing diets are found in geographically limited areas of Indonesia: in Indonesian nutrition literature since prewar times, they have become wellknown as the "Gunung Kidul diet" with dried cassava or gaplek, and the "Seram diet" with sago, as the food staples.

The Gunung Kidul diet is an example of the reverse change from cereals to a non-cereal, a change forced by harsh nature due to deforestation of the uplands and soil erosion, rendering the soil infertile and no longer suitable for cereal cultivation; on the other hand, cassava can still be grown although the yield is below average. This agricultural impoverishment has led to the economic impoverishment of the population (Bailey 1961), mitigated since decades ago by the Government's serious efforts to improve the overall condition of this region through trans-

Table 7. Comparison of the Classical Gunung Kidul and Seram Diets

	Gunung Kidul	Seram
Period of investigation	1938-1939 [a] 1958-1959 [b]	1938-1939 [a] 1969, 1975 [c]
Location	Central Java	Maluku
Population density	High	Low [d]
Environment	Deforested eroded dry limestone hills, infertile soil	Undisturbed swamp and upland forest, fertile soil
Status of inhabitants in terms of food production	Food cultivators or subsistence farmers – "they grow what they eat, and they eat most of what they grow" [b]	Food collectors or harvesters mainly of forest products; fishermen and hunters
Agricultural production	Only cassava can be grown since the turn of this century; formerly rice-maize Legumes (kara) Yield low, below average	Since prehistoric times sagopalms and tubers growing wild in natural stands Fish and game Yield high
Previous staple food	Rice and maize	Sago and tubers since prehistoric times
Present staple food	Dried cassava (tiwul)	Sago and tubers (unchanged)
Proteinrich food	Legumes (kara)	Fish and game
Energy intake	Insufficient [a,b]	Always sufficient [a,c,d]
Protein intake	Quantity: very low [a,b] Quality: poor No animal protein	Quantity: high Quality: good Much animal protein
Evaluation of diet	Poor quality, "poor man's diet" [b]	Good quality
Presence of malnutrition	Severe undernutrition found especially during "paceklik period" (in between two harvest) [b]	No malnutrition found [a]

Source: [a] Postmus and van Veen 1949 [c] Ellen 1977
[b] Bailey 1961 [d] Ohtsuka 1977

migration, soil conservation and rehabilitation, such as terracing and reforestation with lamtoro, Leucaena glauca Benth, irrigation through digging of deep wells, cashewnut cultivation to increase income, and the spread of income-generating projects, etc.

The Seram or sago diet was investigated as well, since sago is also deficient in protein. However, the findings did not conform with the hypothesis that the Gunung Kidul results would be found also in Seram. On the contrary, in general the physical and nutritional status of the population as a whole must be considered good. The explanation is that in the sago diet the energy supply is always sufficient and the protein intake is higher in quantity and of better quality, largely from animal sources such as fish, which is abundant in the coastal areas.

Postmus and van Veen (1949) reported the average daily energy intake in Seram to be about 1,900 Cal/person, which would give an average daily intake for an adult man of 2,280 Cal using the factor 1.2 as man-value, of which 63% is from sago. Ellen (1977) confirmed indirectly the adequate energy supply of the Nuaulu in Seram, reporting a total daily energy intake of 3,085 Cal/adult, of which also 63% (1,958 Cal) was from sago. Ohtsuka (1977), studying the sago eaters in Papua New Guinea, gave the average energy intake of an adult male as 3,600 Cal, of which 65% (2,500 Cal) was from sago, confirming the observation on sago-energy sufficiency, supplemented by the availability of many species of animals, very important sources of protein of good quality.

A brief summary of the differences between Gunung Kidul and Seram may be best given in Table 7. Although the staple foods cassava and sago ambon are nutritionally of the same poor quality, between Gunung Kidul and Seram the differences in the physical environment (ecology) are very striking, with the consequence that a poor man's diet pertains in the former and a well-balanced one in the latter.

FOOD PRODUCTION AND ENVIRONMENTAL STABILITY

One may wonder what kind of changes will occur in the future to the ecosystems of Seram and Irian Jaya through man's intervention, and what the consequences will be amongst others, on the diet and food habit of the population, observing the existing trend from sago to rice. In this context it may be worthwhile to consider Ohtsuka's statement:

> "Also noted is that it is impossible for peoples depending on sago, wild or planted, to migrate out of the low, swampy land, unless they change their staple food. Although the subsistence system dependent on sago is stable if some animal resources are available, the environment is a severely restrictive swamp".

In the existing Indonesian dietary pattern, the use of sago and cassava flours as a sole staple food may have a deleterious effect, as it can be expected that not enough protein of good quality can be obtained from the supplementary foods to meet the protein need. However, as a source of additional energy sago and cassava are most valuable, especially to alleviate the food energy crisis in Indonesia, the same problem faced by so many other developing countries in the world.

As the sagopalm growing in natural stands in the swamp forest dies after it achieves maturation, its starch should be commercially exploited instead of being wasted. However, the preservation of the natural equilibrium of its environment must receive serious attention, to prevent future manmade damage and possible disasters as has happened in the Gunung Kidul area.

Acknowledgement: I gratefully acknowledge the support of the ASEAN Project on Soybean and Protein-rich Foods (ASEAN-Australian Economic Project), Indonesian Institute of Sciences, which awarded me a travel grant to participate in the Second International Sago Symposium.

REFERENCES

AVÉ, J.B. 1977. Sago in insular Southeast Asia: Historical aspects and contemporary use, op. cit., ed. Tan.

BAILEY, K.V. 1961. Rural nutrition studies in Indonesia. I. Background to nutritional problems in the cassava areas, Trop. geog. Med. 13: 224.

ELLEN, F.R. 1977. The place of sago in the subsistence economics of Seram, op. cit., ed. Tan.

FLACH, M. 1977. Yield potential of the sagopalm and its realisation, op. cit., ed. Tan.

FAO/US Dep. of HEW 1972. Food Composition Tables for Use in East Asia; Rome.

HEYNE, K. 1950. De Nuttige Planten van Indonesie (3rd ed.); van Hoeve, s'Gravenhage/Bandung.

INDONESIA, Central Bureau of Statistics, 1977. National Socio-Economic Survey, Fifth Round, Jan-April 1976. Household Expenditure for Consumption (preliminary figures); pub. no. VUS 77-17/18/19; Jakarta.

INDONESIA, Central Bureau of Statistics, 1978. Statistical Year Book of Indonesia 1976. Annual Statistics; pub. no. LUY 77-02, Jakarta: 102-3, 369.

LIE, G.H. 1976. Family food consumption patterns in Indonesia; paper presented at the Symposium on Food and Nutrition, Bogor, 1976: addendum tables 11 and 12 (mimeo.).

LIE, G.H., K.N. Oey, Risnawati Aminah and J. Herlinda 1978. Konsumsi makanan di Indonesia thn. 1976 dibanding dengan target konsumsi dan konsumsi tahun 1970; paper presented at Widya Karya Nasional Pangan dan Gizi, LIPI, Bogor, 1978 (mimeo.).

OCHSE, J.J. 1931. Indische Groenten; Dep. Landbouw, Nijverheid en Handel/Buitenzorg, Volkslectuur, Batavia-Centrum: 283.

OHTSUKA, R. 1977. The sago eaters' adaptation in the Oriomo plateau, Papua New Guinea, op. cit., ed. Tan.

POSTMUS, S. and A.G. van Veen, 1949. Dietary surveys in Java and East-Indonesia I, II, III, Chron. Nat. 105 (12): 317.

SASTRAPRADJA, S. and J.P. Mogea, 1977. Present uses and future development of Metroxylon sagu in Indonesia, op. cit., ed. Tan.

TAN Koonlin (ed.) 1977. Sago-76: Papers of the First International Sago Symposium; Kuala Lumpur.

SAGO PRODUCTION IN SOUTHWEST PENINSULAR MALAYSIA

TAN Koonlin

THE SAGOPALM IN MALAYAN AGRICULTURE

    Aerial reconnaissance for the 1966 and 1974 Present Land
Use Surveys in Malaysia provided unmistakable evidence of a
remarkable resurgence of sago cultivation in the Sungei Batu
Pahat floodplain between those years (Figure 1); statistical
records showed that the area increased from 410 ha in 1960 to
1,746 ha in 1977 (Malaya 1962; Malaysia 1979). A cumulative
acreage of over 11 thousand was planted by peasant farmers in
the state of Johor during 1952-78 (Table 1) under subsidy from
the Rubber Industry Smallholders Development Authority, averaging
nearly 950 acres annually during the decade 1965-74, when large
areas of rubberland (and swamp) were converted to sago palmeries.

Table 1. Cumulative Sago Area in Johor Planted under
RISDA, 1952-1978

District	Acres
Batu Pahat	9,928.50
Pontian	685.75
Muar	422.75
Kluang	164.75
Kota Tinggi	48.25
Segamat	34.50
Mersing	17.25
Johor Bahru	6.00
Johor	11,307.75

Source: Rubber Industry Smallholders
Development Authority, Batu Pahat

*W.R. Stanton and M. Flach (eds.), SAGO. The Equatorial Swamp as a Natural
Resource. Proceedings of the Second International Sago Symposium. All rights reserved.
Copyright © 1980 Martinus Nijhoff Publishers, The Hague/Boston/London.*

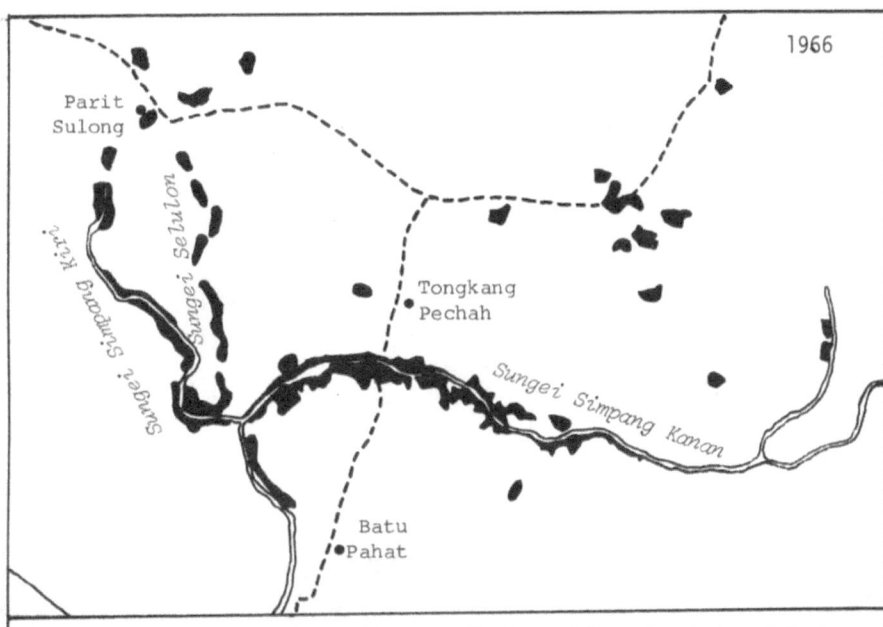

Figure 1.   Expansion of Sago Cultivation in Batu Pahat
District, 1966 and 1974

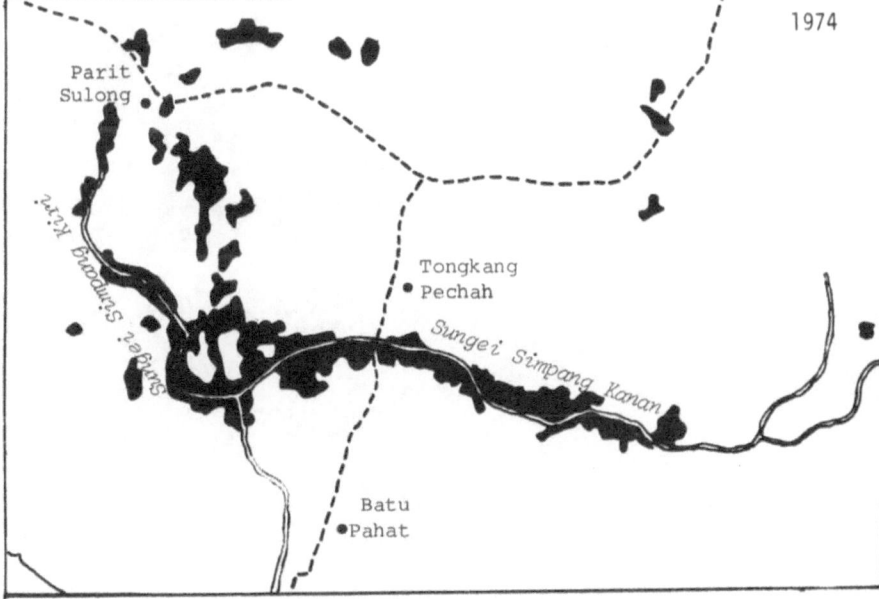

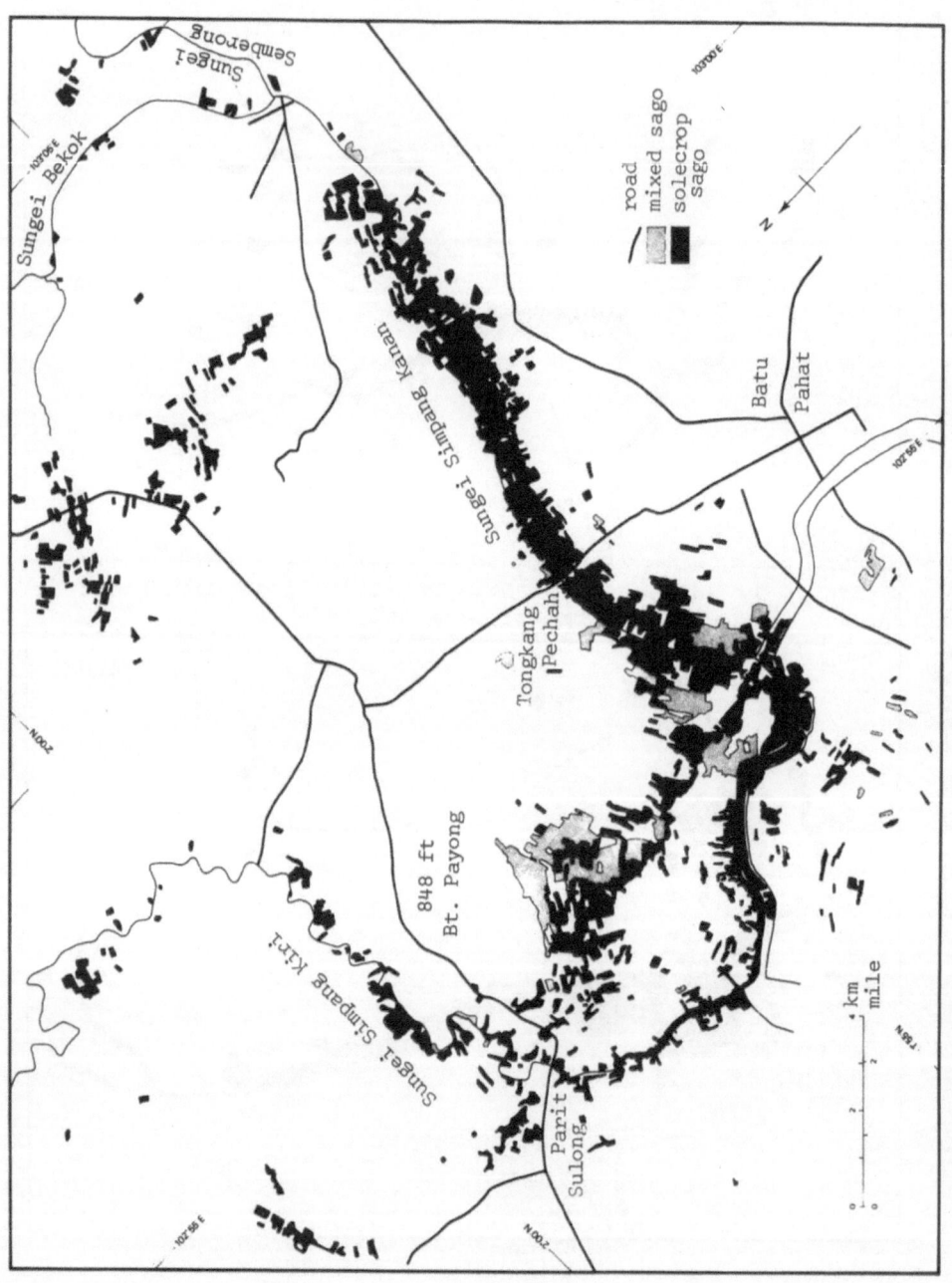

Figure 2. Types of Sago Planting in Batu Pahat District, 1974

The palm, Metroxylon sagu, is cultivated largely in small-holdings as a sole crop by owner-farmers along the tidal banks and humid plain of the twin Simpang tributaries (Figure 2), which rise from the southernmost appendage of the central mountain range, the spring of the major river systems of the south Malayan peneplain. In no other known area in the world is the sagopalm cultivated like a plantation crop as it is in many large holdings here, where the palms are planted at regular distances and subsequently thinned and pruned, and where the harvest is processed entirely for commercial use (Figure 3).

In the evolution of the sagopalm or pokok rumbia as a commercial crop in Malaya, its persistence in this southwest corner of the peninsula appears however to be its last; all earlier centres, notably Melaka (where traditional sago farming over a century earlier (Griffith 1850) had been largely superseded by tapioca, Manihot esculenta), have since become defunct, or are on the verge of extinction, as in Muar and Pontian.

The sagopalm has earned the reputation among the Batu Pahat farmers of having been a crop well worth exploiting within the limitations and opportunities of the freshwater swamp, long before the Ecological Imperative was propounded by modern environmentalists. But its persistence may be likened to a perennial Cinderella in the Malayan agricultural economy; worse, present trends in economic development strategies may well condemn this hitherto useful plant to obscurity for no better reasons than technical expediency and ignorance of its role in environmental wellbeing and the peasant economy and its industrial possibilities.

The myth of the sagopalm as a wild, and by inference also trivial, resource in the peninsula was perpetuated further by transient foreign "experts" engaged to advise on a starch development policy: "In Johore ... the sago palm is not grown as a plantation crop but is found along river banks and scattered in swamp areas by regenerating themselves" (Wahby et al. 1970). This spurious assessment in effect condemned the sagopalm as an object of investment, and denied it a future at a time when ambitious agricultural projects were proliferating in the underdeveloped regions of the peninsula. This was a sad paradox

Figure 3. The Plantation Sagopalm, Metroxylon sagu, in Peninsular Malaysia, to be supplanted by underplanted African oilpalm, Elaeis guineensis. Both palms favour freshwater swamps in their respective native homelands, and have become ecologically competitive on the drained floodplain of the Batu Pahat river. Cocoa, which is a swamp denizen of South America, is gaining popularity as a coconut intercrop in this region. Palms are about 30 ft apart, thinned to solitary stems to facilitate conversion to oilpalm; some sagopalms have been harvested.

against the resurgence of the tapioca industry to meet the rising demand for tropical starches for feedstuffs and fermentation in Europe and Japan earlier so efficiently met by Thailand.

In spite of the fact that sago is still an important secondary cashcrop in Johor state (Table 2), after more than seven decades, however it appears to have been ignored in the latest integrated scheme to develop the western coastal belt of the state, viz. the West Johor Agricultural Development Project. Available reports reveal that the palm was not included in the spectrum of crops selected for extension under Phase 2 of the Project, which encompasses the entire Sungei Batu Pahat basin; even pineapple and smallholder oilpalm are ranked the most important current crops after rubber and coconut, although their areas are smaller than the sago area in this District.

Yet some 35 thousand ha or nearly 10% of the Project area are marshy, and the preliminary intensification of drainage in this region, apparent in the lowered water table, has begun to exert detrimental effects on its ecology, particularly increased acidity of the peaty soils and drought stress in some farmlands

Table 2.   Secondary Crops in Malaya 1977

Crop (ha)	Malaya	Johor
Coconut (smallholdings)	220,324	66,789
Pineapple	14,893	14,589
Cocoa	29,684	3,096
Sugarcane	29,793	3,063
Sago	2,963	2,089*
Tapioca	20,502	1,787
Coffee	8,365	1,594
Pinang	1,783	808
Maize	3,714	417
Nipa	4,879	277
Groundnut	5,960	162

* Including 1,746 ha in Batu Pahat District; officially, sago is known as 'rumbia'.

Source: Malaysia, Min. Agric. 1979

Figure 4. A Neglected Sago Smallholding in the Upper Sungei
Selulon Area, developing into a dense, unkempt grove.

Figure 5. Part of the Plantation Sago Belt on the Sungei Simpang
Kanan, near the confluence; the banks are inundated
daily.

and settlements. The drainage intensity already achieved under decades of peasant endeavours may already have reached the optimum level. Thus current efforts to prepare this erstwhile spongy environment for extensive dryland cropping may turn out in the long term to be misguided and without proper cognisance of environmental principles, especially when this region's rainfall (80 in.) is less than average (120 in.), and it may be incapable of withstanding the consequences of further dehydration if the land is to remain economically productive.

The farmers here have had more than a century of recorded experience with the problems of cultivating the fluvial basin with a succession of orthodox crops, principally rubber and coconuts in this century. It is difficult to understand why so little thought was given to the farmers' unusual reliance on a crop which elsewhere in the peninsula is largely used for thatch, if cultivated at all. With typical peasant pragmatism, they had established mixed holdings in contrast to the common monocultural land use on most floodplains of the peninsula, where their planting skill and investment were spread out over a number of crops; of these the sagopalm used the most marginal habitats, the wetlands, far more efficiently in economic and ecologic terms than any of the other crops they had experimented with.

RISDA helped, as a part of an extensive programme of improving the lot of the peasants, in promoting the cultivation of sago during its postwar rehabilitation as a cashcrop, by including it in their repertoire of alternatives to rubber that the farmers could plant. Farmers had been encouraged to substitute rubber with other crops, as it was recognised that in this basin in particular, the environment was far from suitable for this popular perennial. There is a standard planting grant of $900/acre which in the case of sago is given in instalments of $300 for the first year and $200 in each of the succeeding three years; unlike for most other crops, it barely tides the farmer over the unproductive period of immaturity. The sagopalm now appears to be at the lowest rungs of the crop ladder, no new application for its cultivation having been submitted in the last couple of years.

SAGO TRADE IN THE STRAITS OF MALACCA

Logan made this significant, if not entirely accurate,
observation about the west Johor coastal plain in 1849: "Along
the Sumatra coast from Siak to the Lampongs, and in the large
plains of Rio Formosa and the Muar, there are hundreds of miles
of sago land unoccupied and unproductive, every acre of which
is capable of yielding at the rate of about twenty thousand
pounds of meal yearly". Newbold's (1839) earlier trade compila-
tion showed that these palmeries in the Batu Pahat area most
likely were exploited, because sago was one of the earlier exports
from this area to Singapore; he added: "the trade is carried
on almost entirely by native craft ... and small open boats".

Since it is generally accepted that the sagopalm is not
indigenous to the westerly parts of the Malay archipelago, nor
is it biologically equipped to be an island-hopper, the early
Johor palmeries could not have been wild stands; this assumption
is supported by the fact that the spiny sagopalm, M. rumphii,
is rare in this part of the archipelago, and by the absence of
both species from the wild pandan-rich flora of this swamp
(Corner 1978). It is not known when or by whom the earliest
palms were disseminated to this region, some farmers saying it
spread from Melaka and others that it was brought across the
Straits; their extensiveness intimates long-established if
irregular exploitation.

Logan also assumed that the Malays of southwest Malaya at
the time were not using the palm for their staple food, preferring rice when they could grow or purchase it, and would be
unlikely to have established the palmeries he saw there. But
since there were no other staple foods in abundance in the swamp,
the pioneering farmers probably did use them, as had the Malaccans
theirs earlier. Wheatley (1964) pointed out the fundamental
problem the earliest settlers must have contended with in a virgin
coastal habitat:

> "Despite the wealth and importance of Melaka, the immediate
> hinterland of the port appears to have been very little
> developed .... Muar and Batu Pahat were small farming
> communities .... Few of the immigrants showed any inclina-
> tion to undertake cultivation of the soil. Indeed, farming
> was but poorly developed throughout the south of the peninsula

at this time. Furthermore, the alluvial soils adjacent
to the town /Melaka7 were probably still too saline for
padi-farming. In any case, sago, not rice, was the staple
food of these early colonists, supplemented by quick-growing
crops".

For long "Siak was bound to the fortunes of Melaka, subsequently
Rupat, Jambi and Bengkalis", and it was the Siak traders, operating from that great entrepot city during its eminence, who
imported quantities of the finest sago of the Archipelago from
across the Straits.

While some accounts assert that neither Siak nor Riau produced sago, the indisputable fact remains that they were collecting centres of substantial quantities of sago as well as the
abundant produce of their vast swampy hinterlands for the Melaka
and Singapore trade, where aboriginal serfs produced sago for
export on a sharecropping basis or the surplus was purchased
off aborigines who worked the palmeries for their staple food.
Johor was also part of the Melaka-Riau empires, and by the early
decades of the 17th century, its west coast produced as the only
vegetable foods, a little rice and sago.

A second possible area of influence is Sulawesi, because
rumbia is a Maccarese term for the palm which is used in Sumatra
and in southwest Malaya up to Negeri Sembilan, where an historical
village of the same name exists (Dennys 1894), as also by the
Directorate of National Mapping, Peninsular Malaysia. The Bugis
used to organise substantial prahu trading fleets to Melaka and
later to Singapore; they congregated in Waju in Sulawesi to
collect produce from Maluku (the Spice Islands) and neighbouring
coasts gathered by the aboriginal populations, some of whom were
adept at producing sago bread in exchange for commodities they
lacked.

With the stabilisation of the colonial spheres of exploitation by the mid-19th century and piracy that had disrupted Straits
commerce brought under control, the sago trade from the southwest
Malayan coast resumed, as it also did in neighbouring waters.
Its commercial organisation crystallised in Singapore owing to
the unfavourable trading practices of the Dutch in their own
ports such as Riau, which drove entrepreneurs to develop new

areas of production and trade for the Western markets. "The
Chinese manufacturers in Singapore were the first to introduce
the method of pearling, which has done so much to render it an
article of consumption .... In Singapore all sago is of their
manufacture ... and in Malacca all the sago is of their production"
(Cameron 1865). By the early 19th century they recognised three
grades of sago: the finest came from Siak, next was the Bornean
product, and the least desired was the Moluccan brought over by
the energetic Bugis; as no mention was made of the Malayan product,
it would appear to have been of comparatively inconsequential
proportions.

COMMERCIAL AGRICULTURAL DEVELOPMENT IN JOHOR

The Batu Pahat sago industry is a relict manifestation of
Chinese enterprise in one of many trading activities relinquished
by the maritime Malay merchants and collectors of produce after
trade was internationalised during the colonial era, when hitherto
cumbersome methods of production and simple bartering systems
could no longer meet the burgeoning world demand. But, compared
to the rest of Malaya, Johor came under an occidental system of
commercial agriculture rather late, for the British became involved in the state only in the last years of the 19th century,
who on their arrival noted already flourishing areas of peasant
agriculture on its west coast, especially well organised in the
Muar basin, geared for export mainly to Singapore.

The Johor rulers and the colonial administration encouraged
immigration from neighbouring islands in the late 19th and early
20th centuries to develop the coasts and dilute the Chinese
presence. The waves of immigrants from Melaka, Muar, Sumatra,
Java and Borneo increased particularly during 1895-1911 and
1920-30, to displace the aborigines who were absorbed into the
evolving Malay culture or retreated inland.

These pioneer farmers effected two significant changes to
the once vast, almost impenetrable swamp. They established an
impressive drainage system which now ramifies the southwest
coast of Malaya from Muar to Pontian Kecil for about 15 miles
inland, even deeper into the Sungei Batu Pahat basin. This

drainage in turn enabled the region to be transformed into cultivated fields of a variety of crops, the Bugis being particularly
associated with coconuts in the Benut area. Trading boats pene-
trated past Tongkang Pechah (hence its name), which no longer
do, bringing farm and forest produce and ores to the original
coastal settlement of Bandar Penggaram, core of the present Batu
Pahat town. By 1934, Tempany drew attention to the fact that
Johor possessed the largest holdings of most crops in the peninsula, including sago. Older farmers in the outlying areas of
the District still recall their earlier polycultural landscape.

The Johor Malays already had, with the advent of colonial
rule, the unusual reputation of neither having a wet-rice tradition, preferring where practised dry-rice intercropping on their
mixed holdings especially under coconut, nor developing one even
when encouraged and belaboured by successive administrations,
who perceived an inexplicable prejudice against dry-rice intercropping, even when all their sporadic wet-rice projects of the
interwar era fared dismally, not the least due to their own
ineffectual management of this environment. Most likely the
sagopalm was relied on as a reserve food, as not all could afford
the Javanese rice they relished, for they certainly resorted to
the palm during the Japanese Occupation, unlike most of the rest
of the Malayan population, although this is ignored in most
official studies of the situation of that era.

There is no clear evidence of sago being a major cashcrop
of early British Johor, but certainly its trade was substantial
enough to be recorded regularly from 1911 when the administration
started documenting agricultural progress in the territory; by
1923 (Johor, Ann. Rep.) it was noted: "there is a fairly large
sago palm industry especially in Batu Pahat". This sago, then
produced via the trampling-parut method and known as sagu basah,
is equivalent to the Melanau and Moluccan lementa (Morris 1977;
G.H. Lie 1979, Jakarta, pers. comm.), a raw wet flour after the
bulky fibrous matter, hampas, has been removed through a preliminary wash. Together with similar production from neighbouring
swamplands, sago supported the only notable manufacturing industry
of Singapore in the 19th century; the crude starch was shipped

moist in leafy bags, later gunny sacks, to the port for refining
into starch or pearl for export to Europe, India and America.

GEOGRAPHY OF THE SUNGEI BATU PAHAT BASIN

The Sungei Batu Pahat formed the southern border of the
Portuguese enclave in Malaya after their conquest of Melaka in
1511, and is depicted in Valentyn's map as the Rio Formosa, but
neither the Portuguese nor their successors the Dutch made any
mark on this part of the realm. The river is also famed as the
western connection of the overland route between the Straits of
Malacca and the South China Sea, via the Sungei Semberong, which
joins the Batu Pahat system to the Sungei Endau on the southeast
coast of the peninsula. The Sungei Simpang Kiri, its right
tributary (contrary to occidental custom, the maritime Malays
take their river direction from its mouth, not its source), also
formed the northern boundary of the Orang Binua, distinct from
the Jakun and said to be the most southerly of the Indochinese
aborigines, numerous enough in the mid-19th century to be studied
by a contemporary of Logan but now extinct.

Newbold (1839) characterised the few large rivers of the
peninsula thus: "Their banks are low, swampy, and covered with
mangrove, nipah, nibong, and other trees. Their channels are
for the most part, muddy, except at short distances from the
estuaries". A 1938 map (Arkib Negara) of the riverine environs
of Batu Pahat town shows that this region earlier had the typical
equatorial littoral vegetation: between the mangrove forest
fringe established on the saline tidal flats and the inland
alluvial plain was the swampy basin, built up from the fluvial
sediment of the river, veneered by tidal and floodwater desposi-
tions on which pockets or banks of cultivation were developing;
the intertidal zone was characterised by organic soils, the
Linau-Sedu series.

Poorly drained, the basin was dominated by two ecotones:
the brackish tidal banks supported groves of nipa and belukar,
backed landwards probably up to the hightide mark by permanent
freshwater swamps where undrained, while the drained areas more
likely had a high watertable or were soggy most of the year, on

which commercial sago palmeries subsequently developed. Two
other palms, the arecanut and the coconut were also cultivated
to a large extent, the penchant for palm cultivation indicating
probably some instinctive knowledge of the crop options possible
on this humid habitat, which was reinforced by the domestic value
of these crops, the arecanut supplying thatching poles and a
masticatory, the latter now used largely by the older folks.

The traditional sago area in this basin is centred in the
tongue of land, kuala or muara, at the confluence of the river.
The left bank of the Sungei Simpang Kiri is reputed to be the
geographical core of sago farming, cultivation of the palm here
dating back two centuries according to local lore. The best
palms are claimed to come from this hearth, mula-mula, through
which a sungei mati traverses, the Sungei Selulon, whose meander-
ing has been obscured by silting and drainage works over the
years; it is not clear whether the alleged productiveness of
the palms here is due to the cultivar used, the pedology of the
area, or the undoubted expertise of the established sago farmers.

The kuala was commonly subjected to flood and tidal inunda-
tions; the impeded drainage thus allowed sago to be cultivated
right up to the foothills of Bukit Payong and round its back,
drained by the Simpang Kiri. The water table here is still high,
not often over three feet below the soil in many parts. The
waterlogged nature of the terrain favoured the development of
organic soils, thus rendering it even more unsuitable for many
forms of conventional cropping; the peaty content of this area
can be readily discerned from many drainage channels, the parit,
where the water is perpetually of a tea colour. There is a
circadian cycle of tidal inundation in the lower river banks,
every 25 hours, which decreases in intensity further upstream
and away from the banks, and brackishness is encountered on some
sago holdings under tidal influence. The biology and hydrology
of this basin have yet to be studied comprehensively.

CHARACTER OF SAGO PRODUCTION IN THE BATU PAHAT DISTRICT
Cultivation

The sago industry consists of two sectors, viz. the almost

self-sufficient Chinese planting-processing sector and its Malay
planting adjunct, which appears to be largely Javanese; the latter
derives its commercial survival from the processing sector, which
itself is prey to external market influences. The processing
sector probably controls directly the bulk of the sago area in
the District, for sound economic reasons, but its extent is unknown, as no official data exists on the pattern of sagoland
ownership; it draws its episodic needs for raw materials from
the Malay farms.

The latter group has shown neither willingness nor ability
to participate more actively in the processing and marketing of
their sago produce, and appears content with the prevailing system
of exploitation, for this permits them to pursue other farming
activities, in keeping with their tradition of and still marked
preference for polycultural farming. The biology of the palm,
its agronomic needs, and the mutually beneficial relations of
the two communities promote the almost indifferent Malay attitude
to sago farming, which fits well with the nature of exploitation
of this palm. The superior and unique feature of sago farming
is the baja air, or liquid manuring; the flooding bestows a self-
renewing fertility that no other cropping system can appropriate
or tolerate, bringing returns for the rudimentary drainage-
irrigation system that the farmer establishes on otherwise uncul-
tivable land.

The majority of the Malay sago farmers had ventured into
sago cultivation for commercial gain, apparently emulating the
Chinese, because their land was tanah gambut (peaty), air masin
(brackish), air masam, simpan air (waterlogged), tanah resap
(impermeable), paya (swamp), air pasang (tidal), air tawar (flood-
prone), tanah basah (high water table), tanah rendam and tanah
tenggelam (swampy). More work is needed to differentiate and
assess these different habitats in order to ascertain their
precise ecology, because the farmers provided conflicting information on their suitability for sago growing; the contradictions
may provide clues to the value of various attributes of a habitat
and their combinations for this type of cultivation.

Two-thirds of the sago farmers interviewed for this study

had changed to sago from other crops, especially rubber, either because the marshy land yielded uneconomic returns from dryland crops, or less often, because of long periods of disappointing rubber prices. A sago holding probably epitomises the Malay's ideal of a most agreeable form of investment: a relatively short spurt of hard work leads to a perennial usefulness of marginal property, leaving him much leisure or the alternative of diversifying his likelihood. Not surprisingly, despite the slump, over half the farmers interviewed were reluctant to forego sago cultivation altogether; it should be noted that although Johor has the largest oilpalm area in the peninsula, the peasant farmers in this part of the state are not familiar with this palm.

The system of exploitation developed between the processors and independent farmers enables palms to be sold long before they are mature, and palms are sold not always because they are ready for harvesting but because the farmers may need the income, although naturally the best prices are earned where the palms are allowed to reach maturity. A reliable index of a farmer's financial status is the age at which he disposes of his palms, the wealthier one being more able to nurture his palms for a longer period before the sale, while the poorer one tends to sell more frequently, hence younger palms. During the boom years the processors were prepared to purchase palms from the age of four years onwards, especially when they had attained some two log-heights with their girth discernible (one log = three ft, annual growth; the palm does not manifest perceptible vertical growth until at least two years after planting in the field, longer if not earlier nurtured in a nursery).

Because the Malay farmers do not involve themselves with sago farming beyond nurturing their plantings, their motives for cultivating sago were, with few exceptions, governed not so much by market stimuli as previously by the value of the palm as a reserve food in lean times, by environmental limitations, and by its lack of imposition on their limited resources. The character of sago farming is best summed up by the description, kebun tidor or "sleeping garden" (Johor, Dep. Agric. 1961),

and its attraction by others, kebun simpan, kebun duit, or investment garden, analogies that were not used for their other crops;
these epithets provide an inkling of their farming predilections.

A sago garden then was property that could be "banked" or
"pawned" without loss of the land, unlike in other cropping
systems: the farmer could collect "dividend" by repeatedly
"lending" the inexhaustible "capital", sometimes for years ahead
and to several customers simultaneously, for the polycarpic
sagopalm regenerates itself with little care, after which he
"redeemed" his "capital" intact, the unsaleable palms in turn
providing the next cycle of returns. At worst, the sago farmer
needed not to starve, for the palm in earlier deprived times
had proved that it could provide food and shelter at the least.

A large number of the sago farmers are of the older generation, i.e. over 50 years of age, who were the ones to stress on
the domestic value of the palm. There are two reasons for this
age preponderance: there is a significant volume of outmigration
of young villagers, so that the farmers left behind tend to be
the older ones, who having lived through difficult times also
look upon their sago holdings with favour, even though the present
market fails to justify their faith; in a palmier era they had
been enriched by the sago booms that had enabled them to improve
their homes, invest in more land or in other crops, and even
save for a pilgrimage to Mecca, which the younger farmers had not.

Processing

In the early 1970s, there were over 30 sago factories in
west Johor, the majority concentrated in the coastal belt of
this District; now about half are left, almost all in this
District, centred below Tongkang Pechah along Parit Bilal, their
traditional core (Fairweather and Yap 1937). They include a
few sago-meal factories, a new type of enterprise that taps the
market left void by the demise of the hampas industry, and two
pearl factories, pearling requiring more sophisticated machinery
and far more labour inputs.

All starch and pearl factories require access to abundant
water, although not all have clean water; half a dozen factories

are able to produce a prime-quality starch because they have
access to clear water, usually piped from hill streams, but
because of the prices offered the bulk of production is of medium
quality. The smaller ones which produce sagu basah tend to rely
on parit or river water to rinse the grated mash, and it is
necessary to wash the flour at least twice more before it is
in any fit state for the export market. A few more substantial
processors would have preferred not to buy wet flour, but for
one reason or another they cannot produce enough of their own,
and thus help the smaller processors to stay in business.
Singapore absorbs some Batu Pahat sagu basah and sago for glucose
production.

All sago factories are Chinese-owned, and by the early 1970s
so optimistic had the mood of the industry become that some had
modernised their old plants with suitable components dismantled
from the Singapore sago refineries, which coincidentally had
become anachronistic in the republic's industrial setting. But
the level of technology is not on a par yet with the tapioca
factories in the Chemor area in central Perak, where centrifuging
was already producing a superior grade of starch in that period,
whereas not one of the sago processors appeared to have knowledge
of this method of processing or was interested in using it.
Limited capital resources, lack of interest predisposed by the
market situation and distraction by other more appealing objects
of agricultural investment may be presumed to have caused technological stagnation in the sago industry. In all, it is
estimated that the total dry starch processing capacity does
not exceed 1,000 tons a month were all factories to operate at
maximum efficiency.

The current slump is due to two main factors: there has
been a marked decline in the demand for hampas, so that the
margin of profitability has become proportionately reduced,
aggravated by the dumping of sago on the regional market by
neighbouring states, so that prices have dropped to the lowest
levels. Processors are not encouraged to produce more starch
for low profits and be faced simultaneously with perhaps a cost-
lier waste disposal problem. The lack of a local market for

Figure 7. "Centrifuging", using revolving cylinder covered in fine muslin, in which the grated mash is washed; the starch milk flows out into the trough below.

photo: WRS

Figure 6. Sagolog Pond on Parit Bilal, by the main road below Sungei Simpang Kanan; some 500 logs can be towed by motor boat and punted to the pond.

what was once a coveted pigfeed is due to restrictions on ad hoc
pig rearing, so that smalltime pig farmers who once absorbed the
hampas (which had also been imported from Indonesia often in
large quantities) have practically disappeared from the District;
only poultry rearers use sago hampas as a feedstuff. Further,
because of effective propaganda about the superiority of imported
and cereal based feedstuffs, the more substantial animal farmers
are turning to fortified or complete stockfeeds.

Under the present system of extraction, sago production is
quite labour intensive, and the industry is facing a labour
shortage, particularly in harvesting, because the slump has
prevented processors from competing effectively for skilled
labour; the country's labour shortage is particularly acute in
Johor. Investment in raw materials used to be a considerable
burden, vindicated by the high profits obtained then, but only
a few processors are now able to take advantage of the slump
to purchase stems at the prevailing low prices.

A sago factory always plans at least a year ahead for assured
stocks of stems, but 3-4 years is common practice because it is
difficult to predict the precise period in which a planting can
yield an economic number of mature and nearly-mature palms, as
well as the system here of paying in advance, not at harvest
time as was practised in the Muar area. A factory with an input
of 12 palms a day, with a capacity that can be increased by at
least 50% to meet the highest levels of demand, would therefore
require an annual investment of not less than $20,000 to secure
stocks from independent suppliers, assuming its own holdings can
supply all its basic raw material requirements; over a three-
year period this means a processor must have capital reserves of
some $60,000, assuming an average of $10 per palm at current
price levels for a mature palm, which leaves him in a tight
financial position. In the boom period, when stem prices rose
to $35 per palm, investments in excess of $150,000 were not
unknown. In the absence of centrifuging, sedimenting-settling
and drying space also limits the extra volume of production the
processor may wish to undertake in the event of a favourable
marketing position, even were he able to obtain the palms required.

The 1973 Census of Manufacturing recorded 19 sago/tapioca factories in Johor (presumably the larger of over 30 then listed by the Department of Agriculture, that had a regular labour force), which had a gross output value of over M$2.2 million; all establishments were said to be sago factories, unlike elsewhere in the peninsula (fieldwork in 1973). The price during the 1978 slump fell to about $12 per mature stem (about a tonne each), compared with $33.65/tonne stem in 1973.

PLACE OF SAGO IN THE RURAL ECONOMY

One conclusion I could not help coming to, after a two-month fieldwork in early 1979 in the sago kampongs mostly in the kuala area, was that our agrarian policy does not accommodate the older farmer, who has much to impart in farming expertise as well as manifesting a realistic attitude towards making the optimum if not the most profitable use of his land. In the case of the sagopalm, where no genetic or agronomic studies have been made, in effect commercial palm-starch production is based on a biologically wild plant, their empirical skills acquired through several generations is invaluable for providing a reliable core of knowledge on which scientific studies ought to be generated.

Sago, with its easy agronomy, varied uses and valued role as a source of food and income in the past, was the one crop that could have been promoted for this stratum of the farming community, not only to alleviate poverty in this sector of the population, but also to contribute in a pragmatic if modest way to diversifying the traditional peasant economy and using profitably a skilled labour pool that otherwise remains underexploited. For, unlike younger farmers with growing families and materialistic aspirations, elderly ones have a more conservative lifestyle and take pride in using their skills, hence are unlikely to require land allocations and other support of as great a magnitude as the younger farmers; nor are they easily dissuaded from their endeavours by the temporary miseries of cyclical depressions that characterise export-crop farming. Mixed holdings centred round sago as the main crop on the wetter areas of this District, and as adjuncts to oilpalm and rubber schemes elsewhere that have wetlands, are probably most suitable in all respects for communities of older farmers, presupposing drains (which also serve as transport lines and boundaries in the District) and factories are provided to facilitate commercial exploitation.

Further, the sagopalm, being agronomically flexible, is one of the most useful crops that can be developed extensively to combat the increasing neglect of farmlands because of the outmigration of potential young farmers; some 2 million acres of land are lying fallow in Malaysia largely because of this problem. Mechanisation is being encouraged to reduce the labours of farming in order to attract young farmers, viz. for exacting crops that require such efforts; but the fuel-energy situation has complicated this solution. An alternative approach that ought to be explored is the extension of crops like the sagopalm that are economical in technical and human inputs, particularly when the produce is a raw material with potential as a source of a substitute fuel, agrochemicals, feedstuffs and cellulosic products, viz. capable of generating in situ industrialisation that would also improve rural living by providing employment, amenities and raw materials for manufacturing products that do not require high technological or capital investment.

Starch processing is comparatively simple compared with palm oil, and while it is not recommended to emulate the small group-processing centres for rubber, the starch processors in the peninsula have proved that intermediate prototypes of processing technology already invented are capable of producing good-quality starch and are viable commercially. Where sago-starch is to be used for fermentation and industrial derivatives, manufacturers might be persuaded to accept a grade for non-food uses that could meet modified critical criteria of quality. Also, it may be necessary to formulate a marketing strategy to change the attitude of importers who expect the tropical countries to continue supplying cheap industrial raw materials of high quality that have to compete against subsidised alternatives from industrialised producers, e.g. were the United States to stop subsidising their maize-starch, the tropical starch producers would find the Japanese market less unwilling to accept the better grades

Table 3.   Estimated Sago Group Area in Peninsular Malaysia 1977

	Acreage that could be developed			Acreage already developed(a)	Acreage available for new development(b)
	Suitable	Marginal	Total		
Johor	201,280	547,770	749,050	30,510	720,000
Pahang	127,020	606,460	733,480	61,530	672,000
Selangor	126,920	466,050	592,965	56,210	537,000
Perak	247,430	377,980	625,410	130,200	495,000
Trengganu	144,030	176,650	320,680	84,530	236,000
N. Sembilan	58,070	71,340	129,410	33,320	96,000
Kelantan	29,860	199,420	229,280	186,350	43,000
P. Pinang/S. Perai	31,280	48,700	79,980	44,500	35,000
Kedah	95,700	280,780	376,480	344,160	32,000
Melaka	57,300	-	57,300	31,290	26,000
Perlis	29,920	65,080	95,000	71,190	24,000
Peninsular Malaysia	1,148,810	2,840,230	3,989,040	1,073,790	2,916,000(c)

(a) This acreage includes land under rice and sago only.

(b) A 10% margin of error should be considered.

(c) If the "acreage already developed" were to be deducted from "suitable acreage", one estimate of area available to sago cultivation would be not less than 535 thousand acres.

Source: Chan et al. 1978

of sago-starch at fairer prices than they wish to pay at present.

Sago farming may be considered a model of a fail-safe agricultural system which peasant farmers had developed as the most secure and useful in their precarious environment after several generations of pragmatic labours; lamentably, it appears to have no place in modern agricultural systems. In the aftermath of the Green Revolution of the 1960s, warning against irrational eradication of efficient peasant systems of land use is becoming vocal (Waddell 1977), because modernised systems have usually only succeeded at the expense of social and ecological stability. "Generally speaking, traditional techniques, though limited in terms of productiveness, have the virtue of maintaining the equilibrium between man and his environment, which is more than can be said for modern technologies" (Kassapu 1979). Further, as Harrisson (1970) presaged of similar Malay enterprise in southwest Sarawak: "It is by no means unimaginable that the same conditions of external world-trade break-down which occurred in the Japanese 1940s may re-occur, putting back a premium on local subsistence activities which, in 'better times' are insignificant".

Over a million acres in Malaya are estimated to be suitable for sago cropping (Table 3). Were such habitats in the less developed parts of the peninsula considered for preservation under productive agroforestry, over 500 thousand acres could be placed under sagopalms for energy farming. However, the changing ecology of the region, official obsession with cocoa and the oilpalm as the only viable alternatives to rubber, as well as the slump plaguing the sago market, may at last jeopardise beyond redemption the survival of one of the most durable traditional cashcropping systems in the peninsula.

Like palmsugar and arecanut production, sago could revert to a cottage industry catering to the entrenched local demand for laundry starch and food component. Where before it throve amidst benign official neglect, it is unlikely to survive much longer as a competitive cashcrop in an area where an ambitious rural development programme based on dryland cropping will be vigorously pursued, unless the Government promptly gives it a

proper niche in its agricultural policy or room to develop at
least in its established cores.[a]

It is now asserted that oilpalm cultivation is three times
as profitable, and cocoa even more compared with sago farming;
even rubber has become more profitable to tap. But this was
certainly not so during the 1960s when rubberland was covered
progressively with palmeries. The comparative levels of profitability which reflect adversely on hardy plants like the sagopalm, and are too often used against them, quite conveniently
ignore the enormous capital resources dispensed on draining the
swamps to render them suitable for the cultivation of "cuckoo"
dryland crops and to maintain them effectively afterwards, and
also on the socioeconomic reorganisation of traditional rural
societies. The history of rubber before the Great Depression
(in contrast to its prospects within the next decade) provides
a stark warning that an export crop can be prone to extreme
market forces; nor should it be made even more vulnerable through

---

[a] The oilpalm could become ecologically competitive with the
sagopalm on the drained humid coasts of the peninsula because
the ancestors of the plantation oilpalm of Southeast Asia also
inhabited swamps (of West Africa), and this preference is still
apparent in the well-bred commercial palm.

"The sources and banks of watercourses, moist valleys, especially
those in the rain forest-savanna transition zone, banks of lakes,
swamps, alluvial plains and low-lying islands are the natural
habitats of the palm ..... The oil palm must have been brought
from its natural habitats to its man-induced and the present
habitats, the temporary and permanent clearings in the rainforest" (Zeven 1965). Oilpalm groves also naturalise in habitats
adjoining saline lagoons (Adejuwon 1970).

"Over the last seven or eight years very great improvements of
oil palm on acid sulphate soils have resulted from the suggestion
that the water-table should be raised and maintained as near to
the surface as is practicable" (Joseph and Naruddin Maarof 1977).

ADEJUWON, J.O. 1970. The ecological status of coastal savannas
in Nigeria, J. trop. Geog. 30.
JOSEPH, K.T., and Naruddin bin Maarof 1977. A field study showing
classical changes in acidity over large areas in the Parit Jawa
area of Johor during the last decade, Proc. 3rd ASEAN Soil Conf.
1975, comp. Government of Malaysia; Kuala Lumpur.
ZEVEN, A.C. 1965. The origin of the oil palm (Elaeis guineensis,
Jacq.). A summary, J. Niger. Inst. Oilpalm Res. 4(15).

The wild cocoa too is a swamp habitué.

overproduction, which the oilpalm is in danger of facing at the present rate of conversion of cultivated and forest land to its culture.

Another danger in the prevailing development strategy is that monoculture jeopardises agrarian stability in the long term; moreover promoting one crop at the expense of another useful one is hardly diversification. In the sound management of both our economy and environment the existing pattern of land use ought to be discriminately maintained with due attention to past empirical adaptations that were successful, especially when farinaceous materials are emerging promising fuel-energy resources in an industrialising world. West Malaysian farmers possess a unique pool of skills in plantation-sago agronomy, unlike elsewhere in the ASEAN region where the sagopalm is at best exploited like a semiwild plant.

Acknowledgements: The writer takes this opportunity to express her appreciation to the Stiftung Volkswagenwerk, Federal Republic of Germany and the Institute of Southeast Asian Studies, Republic of Singapore, for enabling her to undertake research on which this precis is based under their first Research Fellowship in Southeast Asian Studies.

REFERENCES

ARKIB NEGARA, Malaysia; Petaling Jaya and Johor Bahru: AA BP 12/20; GA 172/1915; GA 602/1938; Je. 4/J93: Bandar Penggaram (Batu Pahat), 1938 (map scale 16 in. = 1 mile).

BURKILL, I.H. 1935. A Dictionary of the Economic Products of the Malay Peninsula; London: vol. 2.

CAMERON, J. 1865. Our Tropical Possessions in Malayan India; London.

CHAN, Y.K., I.F.T. Wong and W.M. Law 1978. Agricultural development and land utilization, Peninsular Malaysia by year 2000, Proc. Conf. Food and Agriculture Malaysia 2000, ed. H.F. Chin, I.C. Enoch and Wan Mohamed Othman; Fac. Agric., Universiti Pertanian Malaysia, Serdang: 55.

CHEONG, K.C. et al. 1977. Interim Country Report on Malaysia 1977; Tokyo.

CORNER, E.J.H. 1978. The Freshwater Swamp-forest of South Johore and Singapore; Gardens Bull. supp. no. 1, Botanic Gardens, Singapore.

CRAWFURD, J. 1820. History of the Indian Archipelago; London.

CRAWFURD, J. 1856. A Descriptive Dictionary of the Eastern Islands and Adjacent Countries; (rep.) Singapore.

DENNYS, N.B. 1894. A Descriptive Dictionary of British Malaya; London and China Telegraph Off., London.

FAIRWEATHER, J., and S.T. Yap 1937. The sago industry in Malaya, Malay. agric. J. 25.

FLACH, M. 1977. Yield potential of the sagopalm and its realisation, op. cit., ed. K. Tan: 157-80.

FORREST, T. 1780. A voyage to New Guinea and the Moluccas; (repr.) Kuala Lumpur.

GRIFFITH, W. 1850. Palms of British East India (arr. J. McClelland); Calcutta.

HARRISSON, T. 1970. The Malays of South-west Sarawak; London.

JACKSON, R.N. 1961. Immigrant Labour and the Development of Malaya 1786-1920; Gov. Printer, Kuala Lumpur.

KASSAPU, S. 1979. The impact of alien technology: the historical perspective, Ceres 12(1).

LOGAN, J.R. (ed.) 1849. Sago, J. Indian Arch. east. Asia. 3.

MALAYA, Johor. Annual Reports (1911 seq.); Johor Bahru.

MALAYA, Johor, Dep. Agric. 1961. Sago; Johor Bahru (unpub.).

MALAYA, Federation of, 1962. Census of Agriculture 1960. Permanent Crops: Compact Areas and Scattered Trees; Prelim. rep. no. 10; Kuala Lumpur.

MALAYSIA, Ministry of Agriculture, Soils Div. Present Land Use 1966; Present Land Use 1974; Kuala Lumpur.

MALAYSIA, Department of Statistics. Census of Manufacturing Industries, Peninsular Malaysia 1973; Kuala Lumpur: II, 67-71.

MALAYSIA, Ministry of Agriculture, Economics Div. 1979. Area of Miscellaneous Crops 1977; Kuala Lumpur.

MOOR, J.H. 1837. Notices of the Indian Archipelago and Adjacent Countries; London.

MORRIS, H.S. 1977. Melanau sago: 1820-1975, op. cit., ed. Tan.

NEWBOLD, T.J. 1839. Political and Statistical Account of the British Settlements in the Straits of Malacca; London: vol. 1.

SEEMAN, B. 1856. A Popular History of the Palms and their Allies; London.

SINGAPORE 1933-34. Report of the Commission Appointed by His Excellency the Governor of the Straits Settlements to Enquire and Report on the Trade of the Colony (5 vol.).

TAN, K. 1977 (ed.). Sago-76: Papers of the First International Sago Symposium; Kemajuan Kanji, Kuala Lumpur.

TAN, K. 1979. The Economic Potential of the Equatorial Swamp, in terms of Rural Industrialisation and the World's Energy Resources; Institute of Southeast Asian Studies, Singapore (unpub.).

TEMPANY, A. 1934. Report on Agricultural Development in Johore; Johor Bahru.

WADDELL, E. 1977. The return to traditional agriculture, the only means of solving the food problem, The Ecologist 7(4).

WALLACE, A.R. 1898. The Malay Archipelago; (10th ed.) London.

WAHBY, O., M.R. Grace and C.G. Eriksen 1970. The Present Situation of Tapioca Production, Processing and Marketing in Malaysia; Min. Agric., Kuala Lumpur/F.A.O., Rome.

WHEATLEY, P. 1964. Impressions of the Malay Peninsula in Ancient Times; Singapore.

WINSTEDT, R. 1933. A history of Johore (1365-1895), J. Malay. Br. roy. Asiat. Soc. 10(3).

SEED STRUCTURE AND PHYSIOLOGY IN SOME EQUATORIAL PALMS IN
RELATION TO SPREAD AND SURVIVAL

Zaliha Christine ALANG

STRUCTURE OF PALM FRUIT

Whilst many of the families of temperate plants produce one
type of fruit with readily distinquishable characteristic features,
the larger families of tropical trees and woody climbers manifest
a tremendous diversity of size, structure, time of development
and manner of distribution of their fruits or seeds. The Palmae
are one such family (Corner 1966).

Despite their apparent differences, palm fruits have three
features in common: they are fairly large; they usually contain
one large seed; and they do not dehisce.

Figure 1 shows the structure and germination of the fruit
of the oilpalm, Elaeis guineensis, one of the best studied palms.
Most palm fruits develop a pulpy mesocarp at maturity, the texture
of which varies from the softness of the date, Phoenix dactylifera,
and the oiliness of the oilpalm, to the fibrous nature of the
coconut, Cocos nucifera.

Contained within the fruit is the seed. The smallest palm
seeds (about 6 mm long) are found in small undergrowth palms,
whilst the largest is that of the double coconut, Lodoicea
maldivica (about 30 cm). Though generally round in shape, palm
seeds may be somewhat flattened, oblong, conical or even curved.
The outer, hard "shell" of palm seeds is actually the endocarp
of the fruit and is not part of the true seed at all.

Palm seeds consist principally of hard, whitish endosperm,
bounded by a thin, dry, brownish testa, which lies immediately
beneath the shell; within the endosperm the embryo is embedded.
The endosperm is a large mass of triploid, parenchymatous tissue,
which develops more rapidly than the embryo following pollination.

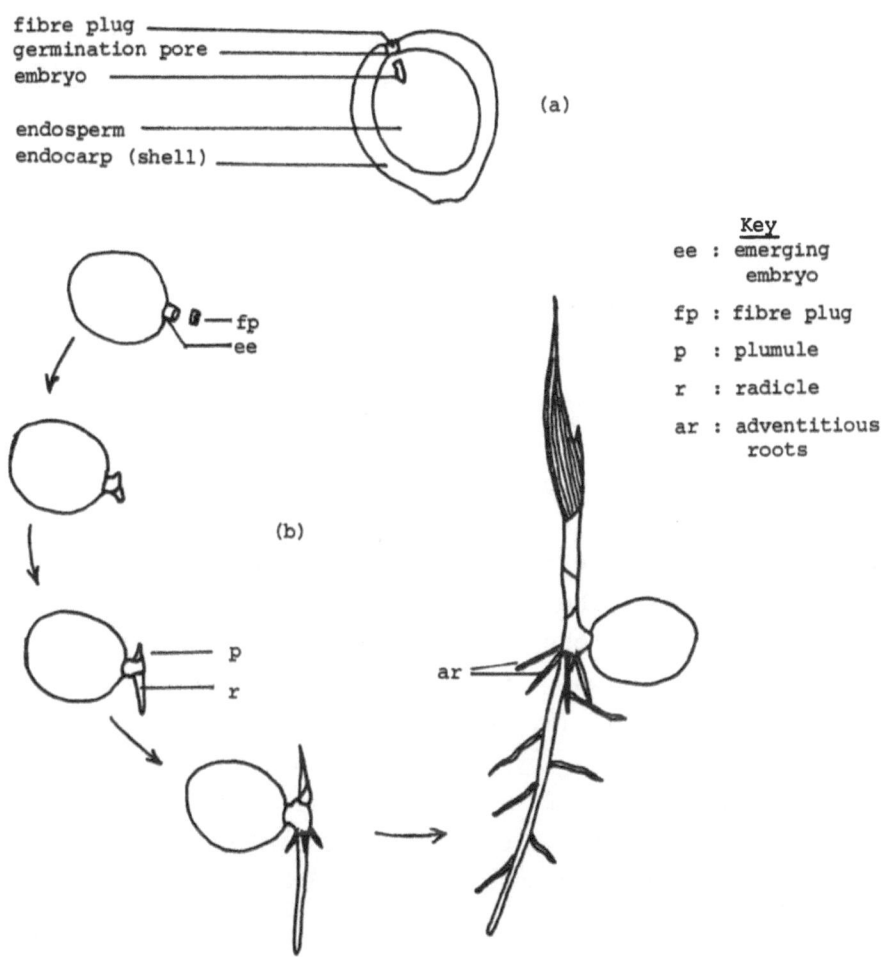

fibre plug
germination pore
embryo

endosperm
endocarp (shell)

(a)

Key
ee : emerging embryo
fp : fibre plug
p : plumule
r : radicle
ar : adventitious roots

fp
ee

(b)

p
r

ar

Figure 1.   Oilpalm Seed and Early Growth of the Seedling
            (adapted from Rees 1960).
            (a) Longitudinal section through an ungerminated seed,
                cut through the embryo;
            (b) Successive stages in germination and early growth
                of the seedling.

It forms a soft, nutritive, semiliquid mass in which the embryo
is embedded. When the seed is almost mature, the walls of the
endosperm cells begin to thicken, from the outer layers of cells
inwards; this thickening of the cell walls, due to the deposition
of a mannan-containing secondary wall which is deposited adjacent
to the primary cell walls, may reach such a degree that the cell

lumens are almost occluded, as in the datepalm (Keusch 1968).

Since the endosperm cell walls are not lignified, the endosperm of palm seeds has a "bony" rather than a "woody" texture; perhaps the best known example is the Ivory Palm nut, Phytelephas macrocarpa, whose endosperm forms "vegetable ivory". The thick cell walls are a form of carbohydrate storage for the embryo, which is solubilised and used during germination. In the coconut, only the outermost region of endosperm solidifies, the central region remaining liquid.

The structure which is often referred to as the "embryo" of palm seeds actually represents the radicle and plumule together with the single cotyledon - all in one apparent "structure". In longitudinal section, the embryonic initials are seen to be oriented such that the radicle points outwards towards the germination pore, whilst the plumule points inwards (Figure 2).

On germination, the embryonic initials are carried out of the seed by elongation of the epicotyl or petiolar region of the "embryo" (Henry 1951) (Figure 1). Under natural conditions, the mesocarp is often eaten by predatory animals and the thin, fibrous shells which are present in some palm "seeds" are easily ruptured, and may even be decaying at the time of germination, due to microbial attack of both the mesocarp and endocarp.

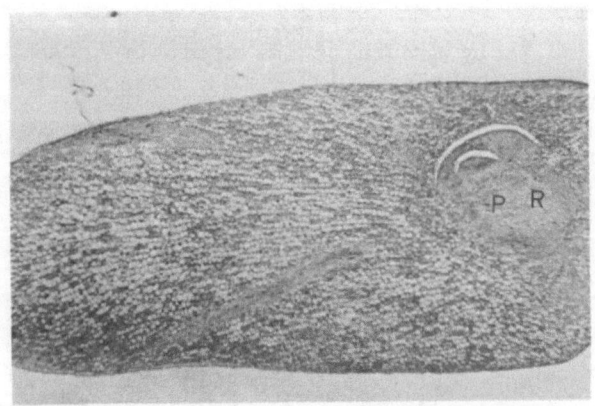

P : plumule
R : radicle

Figure 2. Longitudinal Section through an Oilpalm Embryo (x 40)

In those palms which have a hard shell, there are usually one or more thinner areas in the shell which correspond to the original number of carpels in the ovary of the flower; normally only one of these germination pores is functional, since only one seed develops in most palm fruits. The embryo lies immediately beneath the functional germination pore in most palm seeds, and the embryonic initials are carried out through the pore by elongation of the epicotyl during germination.

The solitary cotyledon remains embedded in the endosperm during the germination process. It grows by cell division and enlargement to form a spongy tissue which eventually fills the entire seed (Figure 3). This spongy cotyledonary structure is called a "haustorium" since it is concerned with the solubilisa-

Figure 3. Stages in the Germination of Oilpalm Kernels. Note extensive development of haustorium (H) at stage 3 ($^2/_3$ natural size).

tion and absorption of nutrients from the endosperm, and the transport of these nutrients to the growing embryo - a function similar to that of the haustorium of parasitic fungi.

The first seedling root (radicle), although strong, is short-lived and is soon replaced by adventitious roots originating at the base of the stem of the seedling. Thus as the stem enlarges, more roots develop, forming a massive network which ramifies through the soil, often at considerable distances from the palm itself.

PHYSIOLOGY OF GERMINATION

Most palm seeds are not dormant; they germinate readily when given the correct balance of moisture, temperature and oxygen. A small, weak, rapidly-growing seedling could not compete with the intense root competition, large amount of debris and low light intensity of their natural environment on the forest floor, but palm seedlings can push their way out of incredibly thick undergrowth and can even establish themselves on rock faces if they have become lodged in a crevice or become entangled in a mass of roots and twigs. This ability to survive is principally due to the large amount of food reserves present in the endosperm of palm seeds and its slow, carefully-controlled degradation by the haustorium.

Thus palms do not need to rely on photosynthesis until their seedlings are fairly well established; this gives them a good start in the low light intensity of their natural habitat - the tropical rainforest. The slow germination of palm fruit is a measure of their massiveness and has undoubtedly contributed to their success as a family.

Flowering and fruit production in the sagopalm (Metroxylon sagu Rottb.) takes place only once at the end of its lifetime, contrasting with the production of fruit in the other commercially important palms, Cocos and Elaeis, which fruit all year round, and Phoenix, which exhibits seasonal fruiting. Sago fruits are straw-coloured when ripe, globose in shape and about 5 cm in diameter, each being covered by 8-9 rows of spiral scales (Kiew 1977; Whitmore 1977). The single seed enclosed within the fruit may be readily germinated (Flach 1977) and although, to the knowledge of the author, no detailed account of germination in this species has yet been recorded in the literature, the development and physiology of germination in the sagopalm is probably similar to that of the datepalm, since both produce carbohydrate-storing seeds.

Although the structure of palm fruits and seeds has been well documented (Corner 1966; Hartley 1977; Vaughan 1970), physiological studies on palm seed germination are few. This is exemplified by the paucity of references to palms in several recent reviews on the subject of germination (Bewley and Black 1977; Mayer and Poljakoff-Mayber 1975; Khan 1977). Studies on the physiology of palm seed germination have been concerned primarily with three economically important species, viz. the coconut-, date- and oilpalm. With a world food crisis now approaching reality, reports of studies on lesser known palms are beginning to appear in the literature (Bovi and Cardoso 1976; Cavalcante 1977).

The literature on the physiology of palm seed germination is rather sparse: the information which is available concerns two main aspects of the subject, one fundamental and the other of more direct agronomic interest. These are:-

1) Use of the endosperm during germination, and the role of the haustorium;

2) Practical methods for controlling the germination of oilpalm seed.

Breakdown and use of the endosperm of palm seeds have received some attention in the case of the oilpalm. The gross chemical composition of the endosperm has been reported and the breakdown of lipid in the endosperm and the absorption of the breakdown products by the haustorium have been investigated (Boatman and Crombie 1958; Opute 1975). Lipids stored in the endosperm are broken down to free fatty acids, which are then absorbed unselectively by the haustorium. The total amount of lipid in the haustorium at any time during its growth is however low compared to the high concentration of lipid in the surrounding endosperm; this is due to a large proportion of the lipid being

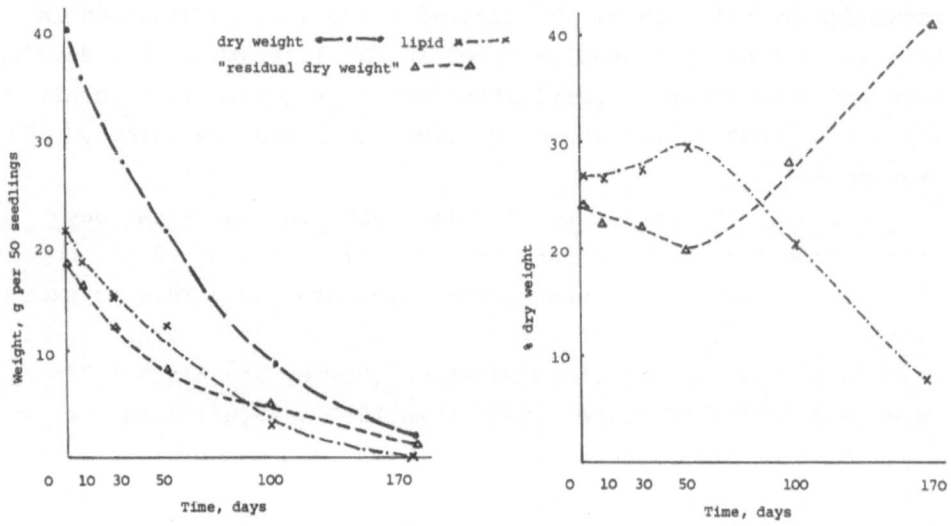

Figure 4. Changes in the Total Dry Weight, Lipid and "Residual Dry Weight" during Germination of Oilpalm Seed. The data (after Boatman and Crombie 1958) refer to the weight of kernels, i.e. endosperm + haustorium + testa.

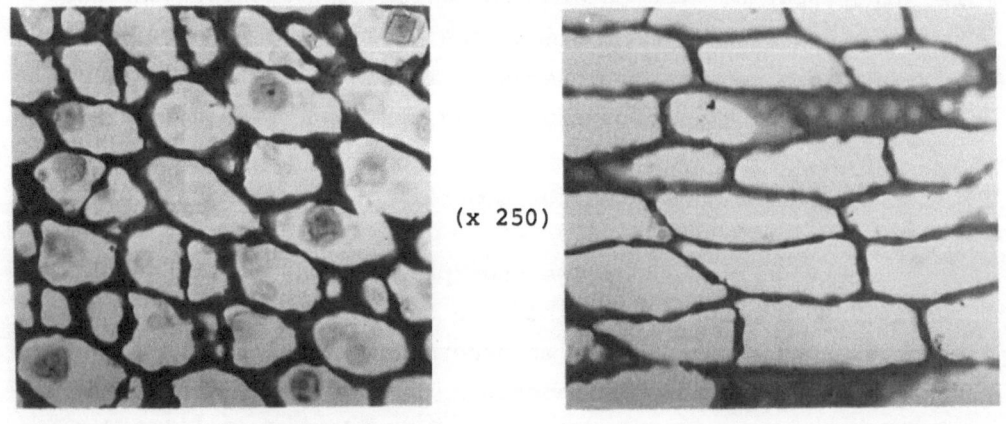

(x 250)

Figure 5. Transverse Section through the Endosperm of an Imbibed,
(left)   Ungerminated Oilpalm Kernel, showing thick cell walls.

Figure 6. Longitudinal Section through the Degraded Endosperm of
(right)   a 6-week Old Germinating Oilpalm Kernel, showing thin residual cell walls.

used as a respiratory substrate for seedling growth, before
photosynthesis begins.

Breakdown of the endosperm takes place in a controlled
manner, with only a small proportion of the endosperm - that
immediately adjacent to the surface of the haustorium - being
softened and degraded at any one time. This suggests that the
activity of the degradative enzymes concerned is under the close
control of the haustorium. The enzymes may be secreted into the
endosperm by the epidermal cells of the haustorium or they may
be activated by a hormonal stimulus originating from the expanding
haustorium. The actual enzymes concerned in oilpalm endosperm,
and the control mechanisms involved, require much further
investigation.

Although the data on oilpalm endosperm degradation show
that the total quantities of endosperm components decrease stead-
ily during germination, calculation of the proportion of the
two major components measured, lipid and "residual dry weight",
shows that the proportion of the former increases slightly during
the initial stages of germination, while the proportion of the
latter decreases slightly (Figure 4; the "residual dry weight"
is mainly insoluble carbohydrate, although some protein and other
substances are included) (Boatman and Crombie 1958). This prefer-
ential use of carbohydrate in the early stages of germination
has been confirmed in a recent study I made (1978, unpub.), and
is considered to be due to the dissolution of the thick cell
walls of the endosperm; this is probably necessary before the
reserve lipid can become available (Figures 5, 6).

In the coconut, the weight of oil per nut does not decrease
during the first 18 weeks after germination (Nathanael 1959),
and it is probable that the mannan present in the endosperm of
coconut is mobilised preferentially (Balasubramaniam et al. 1969),
as in the oilpalm; β-mannosidase activity has been recorded in
the haustorium of the coconut (Balasubramaniam et al. 1973).

In the endosperm of the date palm, mannan forms the major
reserve of the non-oily seed (Keusch 1968). On germination this
reserve carbohydrate is broken down to mannose which is subsequently absorbed by the haustorium. The exact biochemical pathway

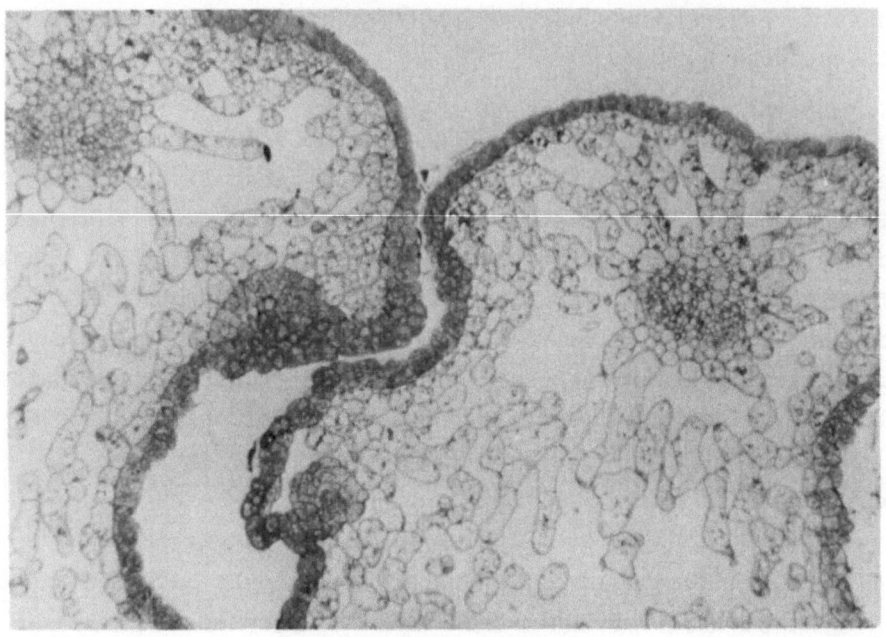

Figure 7. Section through Part of a Haustorium of a 6-week
Old Germinating Oilpalm Seed (x 200)

by which mannose is converted to sucrose within the haustorium
is not known, although a tentative pathway has been suggested.

The importance of the haustorium in palm seed germination
has long been appreciated, yet only relatively recently have
attempts been made to understand the exact physiological role
of this unusual organ. The structure of the haustorium is geared
to the absorption and transport of nutrients from the endosperm
of the developing embryo. Much of the haustorium is composed of
spongy parenchyma which serves to increase the volume of the
haustorium as it grows into the endosperm (Figure 7). The outer
cell layers of the haustorium are more compact and are closely
connected to numerous vascular strands. The surface of the
haustorium is deeply convoluted, increasing the effective area
for absorption.

That the haustorium is a highly metabolic organ is illus-
trated by the wide variety of enzymes, representing many metabolic

Enzymes Detected in the Haustoria of Germinating
Palm Seed

Enzyme	Seed	Reference
Catalase	Oilpalm	Alang (unpub.)
Peroxidase	Oilpalm	Alang (unpub.)
Isocitrate dehydrogenase	Oilpalm	Alang (unpub.)
Glucose-6-phosphate dehydrogenase	Oilpalm	Alang (unpub.)
Isocitrate lyase	Oilpalm	Alang (unpub.)
Amylase	Coconut	Balasubramaniam et al. 1973
Invertase (Sucrase)	Coconut	Balasubramaniam et al. 1973
β mannosidase	Coconut	Balasubramaniam et al. 1973
Phosphatase	Datepalm	Keusch 1968
Urease	Coconut	Nagarajan et al. 1963
Carbonic anhydrase	Coconut	Nagarajan et al. 1963
Ascorbic acid oxidase	Coconut	Nagarajan et al. 1963
Glutaminase	Coconut	Nagarajan et al. 1963
Asparaginase	Coconut	Nagarajan et al. 1963
Aspartate ammonia-lyase (Aspartase)	Coconut	Nagarajan et al. 1963

pathways, which have been identified in the haustoria of various
palms (Table). The exact mechanisms whereby the haustorium
controls the breakdown of lipid and reserve carbohydrate in the
endosperm has not yet been determined in any palm.

Recent work on in vitro culture of oilpalm embryos (Rabéchault and Cas 1974) has shown that the haustorium also has an
important role to play in the growth and development of the
root and shoot initials. Thus it is apparent that there is a
great deal to be learnt regarding the role of the haustorium
during the germination of palm seeds.

COMMERCIAL GERMINATION

The other major topic in palm seed physiology, which has
been investigated in some detail, concerns attempts to obtain
uniform and more rapid germination of oilpalm seeds. Although
the vegetative propagation of oilpalm through tissue culture is

almost a commercial reality (Corley et al. 1976; Rabéchault and Martin 1976), seed production is still the only method currently in use for increasing the numbers of oilpalms for commercial planting. If oilpalm seeds are planted immediately after harvesting or even after some time in storage, germination is sporadic, often extending over periods of up to one year or more. Since this makes the planting out of large numbers of uniformly-developed seedlings almost impossible, a considerable amount of research has been carried out to speed up the germination of oilpalm seeds.

Since the oilpalm embryo itself is not dormant, it appears that some form of mechanical restraint may be responsible for the slow and erratic germination of these seeds. As a result of considerable research (Hussey 1958, 1959; Labro et al. 1964; Rees 1959, 1962), it has been found that oilpalm seeds will germinate rapidly and relatively uniformly if they are heated at 40°C for about two months either in a moist or dry state. Seeds heated under moist conditions tend to germinate during the heat treatment, whereas seeds that have been predried to a moisture content too low for germination will germinate only following rehydration after the heat treatment; this "dry heat treatment" is the method currently in use to produce oilpalm seedlings for large scale planting.

The physical conditions required for the commercial heat treatment of oilpalm seeds possibly resemble the conditions which would occur if the oily mesocarp were broken down by microflora during germination under natural conditions, but the actual physiological processes which take place during the heat treatment have yet to be understood. Initially it was thought that the oxygen supply to the embryo was restricted but later it was shown that oxygen is only effective in stimulating germination after a certain critical period at elevated temperatures; also, when the central region of the oilpalm embryo was exposed to air or oxygen whilst the ends of the embryo remained embedded in the endosperm, no germination occurred (Hussey 1959). Thus there does indeed appear to be a physical restraint to elongation of the embryo.

The region of endosperm immediately above the embryo contains

a ring of smaller, thinner-walled cells which rupture cleanly
on germination, allowing the plug of endosperm and testa to be
pushed ahead of the embryo through the germination pore (Hartley
1977); one might suppose that this process is analogous to abscission. However, application of substances which are known to
accelerate abscission in other plants have so far been ineffective
in promoting the germination of oilpalm seeds (Corley 1976;
Odetola et al. 1975). This apparent ineffectiveness in promoting
germination may be due to poor penetration or incorrect concentration of the chemical substances being used.

There does however appear to be a definite weakening of the
endocarp barrier, and probably of the testa barrier too, following
the heat treatment (Odetola et al. 1975). This, together with
the higher levels of soluble sugars in heat-treated oilpalm
embryos - enabling higher osmotic thrusts - contribute to the
more rapid emergence of heat-treated embryo during germination.
I have detected no major differences in the levels of key enzymes
of several major metabolic pathways in heat-treated oilpalm
embryos, so it is possible that no critical biochemical pathway
is "switched on" during the heat treatment; however, there does
appear to be a higher level of soluble protein in heat-treated
embryos.

In summary, it appears that three complementary processes
are taking place during heat treatment of oilpalm seed:
1) An increase in soluble substances which contribute to
   increased osmotic thrust on imbibition;
2) A weakening of the endocarp and testa barriers; and
3) Accelerated breakdown of the ring of cells in the
   abscission zone of the endosperm.
The latter may be due to the production of, or changes in,
specific levels of hormones which control the enzymes responsible
for cellular breakdown in the abscission zone.

Although the exact physiological mechanism whereby the
commercial heat treatment alters oilpalm seeds so that they
germinate more readily is not yet fully understood, the complexity
of the apparently simple process of germination in the oilpalm
serves not only to illustrate how little is known about the

physiology of palms, two of which are in fact major sources of
edible oil for the world market, but also reflects the success
of the palms as a plant family.

Acknowledgements: I wish to extend my appreciation to
Dr. L.H. Jones, Dr. R.H.V. Corley and Professor G.F.J. Moir
for helpful discussions during the preparation of this paper.

REFERENCES

BALASUBRAMANIAM, K., R.D. Sothary and A.A. Hoover 1969. Hemi-
    cellulose in coconut kernel, Ceylon Assoc. Adv. Sci. 25:
    95 (cited by Balasubramaniam et al. 1973).

BALASUBRAMANIAM, K., T.M.S. Atukorala, S. Wijesundera, A.A.
    Hoover and M.A.T. de Silva 1973. Biochemical changes
    during germination of the coconut (Cocos nucifera), Ann.
    Bot. 37: 439.

BEWLEY, J.D. and M. Black 1977. Physiology and Biochemistry
    of Seeds in Relation to Germination. Vol. 1 - Development,
    germination and growth; Springer-Verlag, Berlin.

BOATMAN, S.G. and W.M. Crombie 1958. Fat metabolism in the
    West African oil palm (Elaeis guineensis) part 2. Fatty
    acid metabolism in the developing seedling, J. exp. Bot.
    9: 52-74.

BOVI, M.L.A. and M. Cardoso 1976. Germination of "acaizeiro
    seeds, Bragantia (Brazil) 35: XCI (cited in Abst. trop.
    Agric. 1977, 19657).

CAVALCANTE, P.B. 1977. Edible palm fruits of the Brazilian
    Amazon, Principes 21: 91 (cited in Abst. trop. Agric. 1978,
    18789).

CORLEY, R.H.V. 1976. Germination and seedling growth, Develop-
    ments in Crop Science I: Oil palm research, ed. R.H.V.
    Corley, J.J. Hardon and B.J. Wood; Elsevier Scient. Pub.,
    Amsterdam: 23.

CORLEY, R.H.V., J.N. Barrett and L.H. Jones 1976. Vegetative
    propagation of oil palm via tissue culture. International
    developments in oil palm, Proc. Malay. Int. Agric. Oil
    Palm Conf.: 1.

CORNER, E.J.H. 1966. The Natural History of Palms; Weidenfeld
    and Nicolson, London.

FLACH, M. 1977. Yield potential of the sagopalm and its
    realisation, Sago-76: Papers of the First International
    Sago Symposium, ed. K. Tan; Kuala Lumpur: 157-77.

HARTLEY, C.W.S. 1977. The Oil Palm (Elaeis guineensis Jacq.);
    (2nd ed.) Longman, London.

HENRY, P. 1951. La germination des graines d'Elaeis, Rev.
    int. Bot. appl. Agric. trop. 31: 349 (cited by Hartley 1977).

HUSSEY, G. 1958. An analysis of the factors controlling the germination of seed of the oil palm Elaeis guineensis (Jacq.), Ann. Bot. n.s. 22: 260.

HUSSEY, G. 1959. The germination of oil palm seed: Experiments with tenera nuts and kernels, J. West Africa Inst. Oil Palm Res. 2: 321.

KEUSCH, L. 1968. Die Mobilisierung des Reservemannans im keimenden Dattalsamen, Planta (Berlin) 78: 321.

KHAN, A.A. 1977. The Physiology and Biochemistry of Seed Dormancy and Germination; North Holland Pub.

KIEW, R. 1977. Taxonomy, ecology and biology of sagopalms in Malaya and Sarawak, Sago-76: Papers of the First International Sago Symposium, ed. K. Tan, Kuala Lumpur: 147-54.

LABRO, M.F., G. Guenin and H. Rabéchault 1964. Essais de levée de dormancé des graines de palmier à huile (Elaeis guineensis Jacq.) par des temperatures elevées, Oléagineux 19: 757.

MAYER, A.M. and A. Poljakoff-Mayber 1975. The Germination of Seeds; (2nd ed.) Pergamon, Oxford.

NAGARAJAN, M., and K.M. Pandalai 1963. Studies on the enzyme activity in the haustorium of the germinating coconut, Indian Coconut J. 17: 25 (cited by Balasubramaniam et al. 1973).

NATHANAEL, W.R.N. 1959. Report of the chemist, Ceylon Coconut Quart. 10: 27 (cited by Balasubramaniam et al. 1973).

ODETOLA, J.A., S.N. Umoessien and T.J. Johnson 1975. Eleventh Ann. Rep., Niger. Inst. Oil Palm Res.: 53.

OPUTE, F.I. 1975. Lipid composition and the role of the haustorium in the young seedling of the West African oil palm, Ann. Bot. 39: 1057.

RABÉCHAULT, H. and S. Cas 1974. Recherches sur la culture in vitro des embryons de palmier à huile (Elaeis guineensis Jacq. var. dura Becc.) X. Culture de segments d'embryons, Oléagineux 29: 73.

RABÉCHAULT, H. and J.P. Martin 1976. Multiplication végétative de palmier à huile (Elaeis guineensis Jacq.) à l'aide de cultures de tissues foliares, C.R. Acad. Sci. Paris 283: 1735.

REES, A.R. 1959. The germination of oil palm seed: the cooling effect, J. West Afric. Inst. Oil Palm Res. 3: 76.

REES, A.R. 1960. Early development of the oil palm seedling, Principes 4: 148 (cited by Hartley 1977).

REES, A.R. 1962. High temperature pretreatment and the germination of seeds of the oil palm Elaeis guineensis (Jacq.), Ann. Bot. n.s. 26: 569.

VAUGHAN, J.G. 1970. The Structure and Utilization of Oil Seeds; Chapaman and Hall, London: 183.

WHITMORE, T.C. 1977. Palms of Malaya; Oxford Univ. Press.

VEGETATIVE PROPAGATION OF PALMS

R.H.V. CORLEY

VEGETATIVE PROPAGATION

In agriculture and horticulture, it is often desirable to have genetically uniform planting material. This can be achieved by using seed from inbred lines, or producing F_1 hybrid seed between such lines. Inbreeding is widely practised with annual crops, e.g. cereals, but with cross-pollinating perennials an inbreeding programme usually requires too great a time to be useful. An alternative method of obtaining uniform planting material is by vegetative propagation, traditionally by grafting, taking cuttings, or using bulbs or tubers; many horticultural and perennial tree crops are propagated by such methods. The possibility of using tissue culture for vegetative propagation is also being investigated for several crops and is already used, on a commercial scale, for propagation of some ornamental plants.

Many palms, e.g. Metroxylon, Phoenix, produce suckers from the base of the stem; these suckers can be removed and rooted and thus used to propagate the palm. However, the number of suckers produced is usually small - Oudejans (1969) mentions 5-25 per tree for Phoenix dactylifera - so the rate of multiplication obtainable is very low. Some economically important species, e.g. Cocos nucifera, Elaeis guineensis and the related E. oleifera, do not produce suckers, and hitherto vegetative propagation of these species has been impossible.

The use of tissue culture methods for propagation of palms has been under investigation for several years, and encouraging results are now being obtained. With E. guineensis, propagation methods have been developed (Jones 1974; Rabéchault and Martin 1976), and are currently being used, on a pilot scale, for

W.R. Stanton and M. Flach (eds.), SAGO. The Equatorial Swamp as a Natural Resource. Proceedings of the Second International Sago Symposium. All rights reserved. Copyright © 1980 Martinus Nijhoff Publishers, The Hague/Boston/London.

production of planting material (Corley et al. 1977, 1979).

Plants have been produced from cultures of Phoenix dactylifera and other palms (Ammar and Benbadis 1977; Reuveni 1979; Reynolds and Murashige 1979), while tissue cultures have been established from C. nucifera (Eeuwens 1978; Fisher and Tsai 1978; W.K. Smith, pers. comm.).

Tissue Culture Methods

The tissue culture process, as applied to palms, involves several distinct stages.

a) Disinfestation. Pieces of suitable tissue (explants) are removed from the chosen parent palm, and treated with, for example, sodium hypochlorite solution to destroy fungal spores and bacteria. This disinfestation is necessary because fungi and bacteria grow much faster on the culture medium than do the explants themselves. The sterile explants are placed on, or in, a sterile medium containing mineral salts, vitamins, an energy source (usually sucrose) and plant hormones to regulate growth. With the oilpalm, suitable tissues include the apical meristem, young leaves, roots and immature inflorescences. However, apical meristem cannot be used without destroying the palm while, if reproductive structures are used, care must be taken to ensure that cultures are initiated from maternal cells and not from cells that have undergone meiosis.

b) Callus initiation and growth. Given the right culture conditions (growth regulator concentrations are usually critical), callus will develop from the explant. Callus consists of a disorganised mass of growing and dividing cells and, once initiated, it can often be persuaded to continue growing almost indefinitely; thus large quantities of callus can be obtained from a single explant. This potentially unlimited growth allows the amplification which is so desirable in a propagation method.

c) Plant regeneration. By changing the culture conditions (again, growth regulator concentrations are usually important), shoots, roots or embryo-like structures called embryoids can be differentiated from the callus. With oilpalm, if roots are allowed to develop first, we have found it impossible to produce shoots later; however, if shoots develop first, later initiation of adventitious roots presents few problems. Balanced embryoids, which produce root and shoot simultaneously, have been comparatively rare in our oilpalm cultures.

Plant regeneration appears the most difficult step in the propagation process at present. With oilpalm it is not yet possible with all genotypes. With coconut, callus initiation is fairly easy, but we have not yet succeeded in regenerating plants (Smith, pers. comm.). Blake and Eeuwens (1978) have obtained shoot-like structures from inflorescence explants, but there appears to be no callus stage in their method. With Phoenix, regeneration has only been achieved from callus derived from seed embryos (Reuveni 1979; Reynolds and Murashige 1979).

d) Transplanting. Once well-rooted shoots have developed in culture, they can be transplanted to soil. Some hardening is usually needed, as the leaves produced in culture often fail to develop a normal cuticle (e.g. Wardle et al. 1979), and are easily dessicated. Hardening may involve: (i) exposure of the plant to a dry atmosphere before transplanting, e.g. by removing the cap of the culture tube; (ii) maintenance of high humidity after transplanting, until new leaves are produced or the root system is well established; and (iii) use of a polyvinyl antitranspirant film, to replace the missing cuticular wax. In our experience with oilpalm, the most effective procedure has involved maintaining a high humidity for 3-4 weeks after transplanting.

e) Meristem proliferation. With many plant species, the callus phase can be avoided, by explanting meristem (buds) and inducing them to proliferate. This has advantages because with many plants chromosomal abnormalities tend to develop in callus cultures, and a proportion of the regenerated plants may then be abnormal. With oilpalm, attempts to induce meristem proliferation have not yet been successful; however, it is worth noting that our results so far indicate that very few abnormalities develop in oilpalm cultures. Possibly palm cells, which can be extremely long lived (Parthasarathy and Tomlinson 1967), may have a greater inherent chromosonal stability than the shorter lived cells of dicotyledons and herbaceous monocotyledon species.

RESULTS WITH OILPALM

Three plants of the first oilpalm clone produced by tissue
culture were planted in the field in January 1977. Since then,
a total of over 900 plants, representing 20 different clones,
have been planted out, and many more are established in the
nursery.

Harvesting of the three palms of the first clone has com-
menced, and by 30 months after field planting (at which time
harvesting of oilpalms normally starts), they had already yielded
over 30 kg of fruit per palm, or the equivalent of 0.6 t/ha
oil. The oil content of the first bunches was low (this is
also true for ordinary seedlings), but the most recently harvested
bunches had over 25% oil to bunch weight, which is quite acceptable. Viability of pollen from the palms of this clone exceeded
90%; seeds from the first bunch have been germinated, and
the seedlings appear quite normal.

Table 1. Leaf Production of Clone JB 32 and Seedlings

Planting material	Leaf production per palm per 6 months	Coefficient of variation %
Clone JB 32	12.1	8.9
Mixed seedlings	9.8	11.6
Progeny 1	10.0	14.2
Progeny 2	9.8	11.0
Progeny 3	10.6	10.8

Corley et al. 1979

Early results from more extensive plantings of other clones
show the greater uniformity of clonal material, relative to
seedlings; this was also observed in the nursery (Corley et al.
1977). Tables 1 and 2 show that, with the exception of Clone
20.1, the variation between clonal palms in rate of leaf production is less than that between seedlings. Clone 20.1 shows
a tendency to suffer greater transplanting shock than average,

Table 2. Leaf Production of Five Clones and Seedlings

Planting material	Leaf production per palm per 7 months	Coefficient of variation %
Clone JB 12.26	12.6	5.6
Clone 171.97	10.8	8.5
Clone JB 20.1	12.4	15.1
Clone JB 20.49	14.3	7.2
Clone JB 22.24	14.7	7.0
Mixed seedlings	10.6	14.2

Source: Corley et al. 1979

so the variable leaf production in the first seven months after
transplanting may be due to varying degrees of shock.

Table 3 shows that Clone JB 32 is less variable, in terms
of the leaf number subtending the first inflorescence, than
seedlings. The higher mean leaf number for the clone is a
reflection of the higher rate of leaf production observed (Table
1), and is not due to any difference in the time of onset of
flowering (last column, Table 3).

Table 3. Early Flowering of Clone JB 32 and Seedlings

Planting material	Number of leaf bearing first inflorescence*	Coefficient of variation %	Palms flowering 1 yr after planting %
Clone JB 32	42.1	8	72
Mixed seedlings	33.6	24	45
Progeny 1	31.6	16	70
Progeny 2	30.0	19	52
Progeny 3	29.7	16	82

* Counting the first true leaf, in the nursery, as leaf 1

Source: Corley et al. 1979

When individual palms of a clone are compared, the uniformity
of flowering pattern is quite remarkable. Table 4 shows the
sequence of male and female inflorescence production for palms

Table 4. Sex of Inflorescences in Successive Leaf Axils of Clonal and Seedling Oilpalms in the Same Environment

Palm number	Clone JB 20.32			Seedlings			
Leaf axil:*	1	2	3	1	2	3	4
1	M	M	M		M	M	M
2	M	M	M		M	M	M
3	M	M	M	F	M	M	M
4	M	H	M	F	M		M
5	H	H	M	M	M	F	M
6	H	F	H	M		F	
7	F	F	H	F	M	F	
8	F	F	F	M	F	F	M
9	F	F	F	M	F	F	
10	F	H	F	F	F	H	F
11	F	F	H	F	F	F	F
12	F	F	F	M	F	F	M
13	F	F	F	F	F	F	M
14	F	F	F	M	F	F	M
15	F	F	F	M	F	F	
16	F	F	F	M	F	F	F
17	F	F	F	M	F	F	F
18	F	F	F	M	F	F	F
19	F		F	M	F	F	F
20	F			F	F	F	
21		M	M	F	F	F	M
22	M	M	M	F	F	F	F
23	M	M	M	M		F	F
24	M	M	M	M		F	F
25	M	M	M	F	F	F	F
26	F	F	F			F	F
27	F	F				F	F
28	M	M	M			F	F
29	M	M	M			F	F
30	M	M	M	M	F	F	F

* M = male; F = female; H = mixed male and female spikelets on the same inflorescence

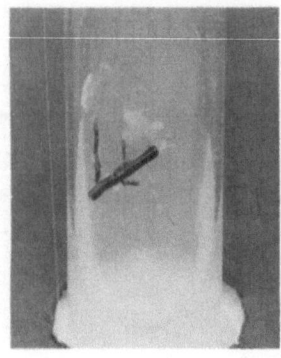 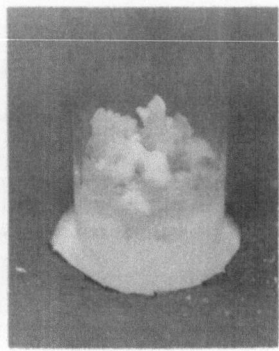 

Propagation of Oilpalm by Tissue Culture

(from upper left) Root explant with callus
Callus with embryoids
Small shoots developing from embryoids

(from lower left) Rooted shoots, ready for transplanting to soil
Plantlet established in soil
One-year old nursery palm

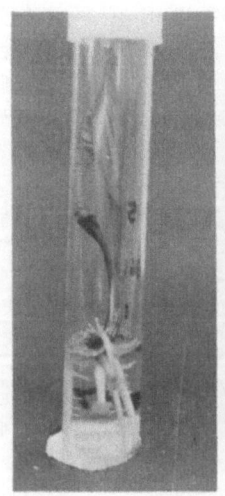  

(left) Young oilpalm, two years after transplanting from culture
(right) Developing fruit bunches, three years after transplanting
        from culture

of a single clone, and for seedlings in the same environment
(oilpalms produce separate male and female inflorescences on
the same palm). For oilpalm, the uniformity of flowering behaviour may be a disadvantage since, to ensure adequate pollination
and fruit set, some palms in a field should produce male inflorescences while others are in a female phase. This potential
problem might be overcome by planting mixtures of two or more
clones. With sago, uniformity of flowering behaviour might be
an advantage, in that the optimum time for harvesting might be
more easily predicted.

The early performance of these oilpalm clones is therefore
encouraging. However, as noted elsewhere (Corley et al. 1977),
the first clones, produced in a Unilever Research Laboratory in
Britain, are all derived from unselected seedlings, and on average
their performance is unlikely to be any better than that of
normal seedlings. A few clones may prove good enough for commercial planting, but extensive field testing will be required
to identify such clones with certainty. Clearly, it is preferable
to propagate from selected mature palms with known performance.

With this objective in view, we have so far initiated callus cultures from over 150 selected mature palms. Some palms have proved much more difficult to culture than others, and the growth rate of some callus lines is low, but we have succeeded in starting cultures from all the palms attempted. Most clones have doubling times, in culture, in the region of 20 days; such a doubling time allows production of 300,000 cultures within a year, starting from a single culture. Since from each culture we may regenerate several plantlets, it is clear that production of commercial quantities of planting material will be quite feasible.

DISCUSSION

Our work with oilpalm indicates that tissue culture can be used for vegetative propagation of palms on a commercial scale. It is worth considering therefore, the advantages of this technique for some of the economically useful palm species.

With the oilpalm, E. guineensis, vegetative propagation has not been possible before, and the crop has been grown entirely from seed. Seedling progenies are variable, and even the best progenies contain some poor palms. Vegetative propagation of the best palms from the best progenies should allow yields at least 30-50% above current levels to be obtained. The cost of producing clonal plants by our present tissue culture methods will be much higher than the cost of producing seedlings, but the extra cost of planting material will be easily justified by the higher yield expected.

Vegetative propagation will also allow the oilpalm breeder to start selection for such "secondary" characteristics as disease resistance, improved oil composition, short trunks (for ease of harvesting), efficient fertiliser usage, and so on. Where palms are grown from seed, it is necessary to identify male and female parents, which when crossed will give progenies with both high yield and the required secondary characteristics. With vegetative propagation, on the other hand, it is only necessary to identify a single palm with the desired genetic constitution, and clonal planting material can be produced from it.

The coconut is also grown from seed, and propagation by tissue culture will have the same advantages as with oilpalm. However, because of the small number of nuts produced per palm, the cost of coconut seed production is as high as the likely cost of the tissue culture method. At present the supply of Dwarf x Tall hybrid seeds is inadequate, and tissue culture propagation might profitably be used to supplement this supply. In the longer term, of course, the production of elite clones must be the objective, but even unimproved clones could be worth producing at present.

With the date palm, P. dactylifera, a method of vegetative propagation already exists, and indeed all commercial date culti-vars are propagated vegetatively. However, the present method of propagation is slow and tissue culture might well be compe-titive, in terms of cost. Date breeding programmes are compli-cated by the fact that the palm being dioecious, all commercial cultivars consist exclusively of female trees (Oudejans 1969). Thus direct intervarietal crosses are impossible, while crossing the highly selected female cultivars with unselected males will generally give poor quality progenies. Such progenies might contain some outstanding individual palms however, and tissue culture propagation would allow a rapid multiplication and exploitation of such trees, which is not possible with the conventional propagation method.

With Metroxylon, vegetative propagation is the normal method of propagation but, as with the date palm, the production of suckers is limited. Breeding and selection are complicated by the fact that, for a proper assessment, the tree must be destroyed in harvesting, before it flowers. Suckers from the best trees can be selected, thus gradually increasing the proportion of good clones, and these suckers could be bred from, albeit a generation later than the selection was done. Alternatively, a non-destructive method of assessing starch content, from trunk core samples, would allow selection followed by breeding from the same tree. (Controlled pollination could be difficult because male and hermaphrodite, but no separate female flowers are produced (Tomlinson, 1971), in contrast to Cocos, Elaeis, or Phoenix, which have female flowers, inflorescences, or trees, respectively).

Because assessment must be based on a single harvest, more extensive replication would be needed in sago clone or progeny trials than with the fruit-crop palms, where yields can be recorded over several years. Tissue culture would allow propagation on a scale adequate for proper clone trials, as well as permitting rapid exploitation of any outstanding clones identified. It appears likely that, given some breeding effort and a rapid method of propagation, good sago clones could become a highly productive source of starch.

Barrau (1959) quotes yields of 7-9 $t.ha^{-1}.yr^{-1}$, with a water content of 35-40%; this is equivalent to only 4-6 tonnes of dry starch. However, individual palms may yield over 500 kg (Burkill 1966); if 60 such trunks could be harvested per ha per year (Barrau 1959), 30 t of starch might be obtained. Flach (1977) mentions a yield of 25 t of starch per ha has being achieved in the Batu Pahat District in Johor. These figures compare well with the highest yields recorded from cassava, or from multiple-cropping of rice.

There are many other palm species of economic importance, e.g. Areca catechu, the betelnut palm; Copernicia cerifera, the carnauba wax palm; Guilielma gasipaes, the peach palm; Nypa fruticans; Orbignya spp., the babasu palms; Raphia spp. With all of these, vegetative propagation is either slow or impossible, and there can be little doubt that the tissue culture technique developed for oilpalm could have a major impact on most other palm crops.

Acknowledgements: I am grateful to Unilever Ltd. for permission to publish.

REFERENCES

AMMAR, S., and A. Benbadis 1977. Multiplication végétative du palmier dattier (Phoenix dactylifera L.) par la cultures de tissus de jeunes plantes issues de semis, C.R. Acad. Sci. Paris, 284 ser. D: 1789-92.

BARRAU, J. 1959. The sago palms and other food plants of marsh dwellers in the South Pacific Islands, Econ. Bot. 13: 151-62.

BLAKE, J., and C.J. Eeuwens 1978. Inflorescence tissue as source material for vegetative propagation of the coconut palm; paper presented to 1978 International Conference on Cocoa and Coconuts, Kuala Lumpur.

BURKILL, I.H. 1966. A Dictionary of the Economic Products of the Malay Peninsula; (2 vol.) Min. Agric. Coop., Kuala Lumpur: 2444 pp.

CORLEY, R.H.V., J.N. Barrett and L.H. Jones 1977. Vegetative propagation of oil palm via tissue culture, Oil Palm News 22: 2-7.

CORLEY, R.H.V., K.C. Wooi and C.Y. Wong 1979. Progress with vegetative propagation of oil palm, The Planter, Kuala Lumpur (in press).

EEUWENS, C.J. 1978. Effects of organic nutrients and hormones on growth and development of tissue explants from coconut (Cocos nucifera) and date (Phoenix dactylifera) palms cultured in vitro, Physiol. Plant. 42: 173-78.

FISCHER, J.B., and J.H. Tsai 1978. In vitro growth of embryos and callus of coconut palm, In Vitro 14: 307-11.

FLACH, M. 1977. Yield potential of the sago palm and its realisation, Sago-76, ed. K. Tan, Kemajuan Kanji, Kuala Lumpur: 157-77.

JONES, L.H. 1974. Propagation of clonal oil palm by tissue culture, Oil Palm News 17: 1-8.

OUDEJANS, J.H.M. 1969. Date palm (Phoenix dactylifera L.), Outlines of Perennial Crop Breeding in the Tropics, ed. F.P. Ferwerda and F. Wit; Wageningen, misc. pap. 4: 243-57.

PARTHASARATHY, M.V., and P.B. Tomlinson 1967. Anatomical features of metaphloem in stems of Sabal, Cocos and two other palms, Amer. J. Bot. 54: 1143-51.

RABÉCHAULT, H., and J.P. Martin 1976. Multiplication végétative du palmier a huile (Elaeis guineensis Jacq.) a l'aide de cultures de tissus foliaires, C.R. Acad. Sci. Paris, 283 Ser. D: 1735-7.

REUVENI, O. 1979. Embryogenesis and plantlets growth of date palm (Phoenix dactylifera L.) derived from callus tissue, Hort. Sci. 14: 457-8.

REYNOLDS, J.F., and T. Murashige 1979. Asexual embryogenesis in callus cultures of palms, In Vitro 15: 383-7.

TOMLINSON, P.B. 1971. Flowering in Metroxylon (the sago palm), Principes 15: 49-62.

WARDLE, K., A. Quinlan and I. Simpkins 1979. Abscissic acid and the regulation of water loss in plantlets of Brassica oleracea L. var. botrytis regenerated through apical meristem culture, Ann. Bot. 43: 745-52.

SAGO PALMS FROM EQUATORIAL SWAMPS;
A COMPETITIVE SOURCE OF TROPICAL STARCH

M. FLACH

THE MAIN MOISTURE-RICH STARCHY STAPLES

Production and Productivity

The main moisture-rich starchy staple foods are listed in
Table 1, together with their approximate chemical composition;
it is clear that these foodcrops to a certain extent are mutually
exchangeable. This especially holds true from the point of view
of starch production in which a high protein content would be
a disadvantage.

The distribution of the production of the crops over the
world is presented in Table 2. In mainly tropical and subtropical
countries the diversity of starch crops is greater than in the
rest of the world. Production per head of population in non-
tropical countries, however, is some 50% higher than in the
tropical countries.

A comparison of productivity of the crops is given in Table 3
and also presented in Figure 1. The average world yields in 1963
and the experimental top yields until 1967 were taken from de
Vries et al. (1967). Average 1974 world yields were derived from
FAO (1976); sources for experimental top yields are also given.

The average productivity of the sweet potato is rather high,
if compared to other crops. This probably is due to its cultivation in Japan and the USA; however, its highest productivity is
obtained in Nigeria. The potato is, due to its being cultivated
mainly in non-tropical countries, excluded from consideration.
The sagopalm, which until now did not, or hardly, received attention in research, in this comparison appears to do very well.

Only two of the crops are perennials, i.e. plantain and
sagopalm; the others are short duration crops. This means that,
due to tillage requirements, the yields of the short duration

Table 1. Comparison of Composition of the Main
Moisture-rich Starchy Staple Foods

Composition	Cassava	Sweet potato	Aroids	Yam	Potato	Sagopalm	Plantain	Rice
% edible of fresh product	83	88	85	85	88	–	59	70
KCal per 100 g of fresh product	153	114	113	104	75	76	128	350
Dry matter content	40	30	30	27	20	20	33	85
% carbohydrates on DM	92.5	85.8	85.8	88.8	85.0	92.5	93.0	88.6
% protein on DM	1.8	5.0	6.6	7.4	10.0	1.5	3.0	8.0
% fat on DM	0.5	1.0	Ø	0.7	Ø	0.5	0.6	1.7
% fibre on DM	2.5	3.3	1.7	1.9	2	1.5	0.9	0.2

Note: All figures after Platt (1962) except for sagopalm, for
which the figures of Lim (1967) on pith were used.

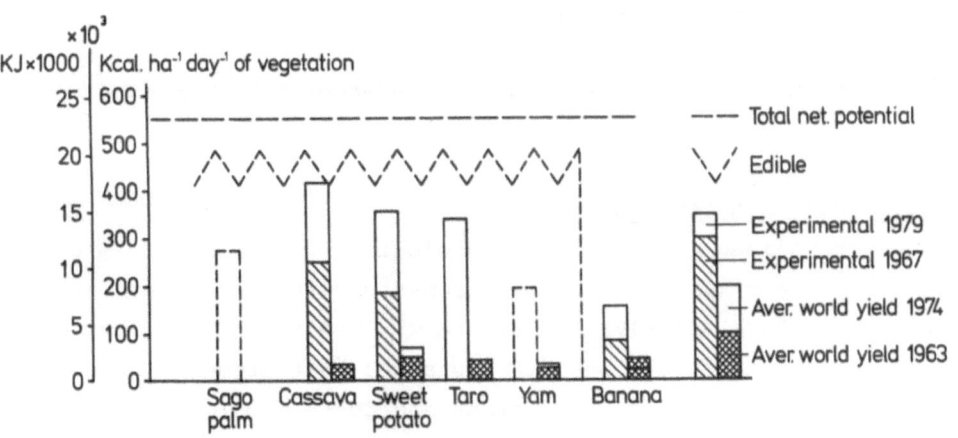

Figure 1. Average World Yields of Selected Staples, 1963 and 1974.
Maximum yields obtained in experiments until 1967 and
1979 compared with potential production of both total
dry matter and edible dry matter in KJ (Kcal) per ha and
day of vegetation; for yam an estimate is given, and for
sagopalm an experienced able farmer's yield is given.

Table 2. Distribution of Production of the Main Starchy Staples over the Main Climatic Zones, in relation to the Population of these Zones

| Crop | Total Production | | | % of Total Production | | | |
| | | | | Mainly tropical and subtrop. countries | | Other countries | |
	10^6t	4.2×10^{12}KJ or 10^{12}Kcal	% of energy	weight of crop	energy of total	weight of crop	energy of total
Potato	134	208.4	40.7	4.5	1.8	95.5	38.9
Sweet potato	134	134.4	26.3	12.8	3.4	87.2	22.9
Cassava	103	130.8	25.6	99.3	25.4	0.7	0.2
Yam	20	17.7	3.5	99.3	3.5	0.7	0.0
Aroids	4	3.8	0.7	89.6	0.6	10.4	0.1
Plantain	18	13.9	2.7	100	2.7	0.0	0.0
Sagopalm*	3	2.4	0.5	100	0.5	0.0	0.0
Total	579	511.4	100	-	37.9	-	62.1
Population	4.10^9			1.9×10^9		2.1×10^9	

* Own rough estimate

Source: Calculated from FAO (1976) statistics

crops in the comparison are somewhat exaggerated, the more so the shorter the lifespan of the crop.

Ecological Requirements

Generalised information on ecological requirements is compiled by Flach (1980). The requirements with respect to water and temperature are given in Figure 2. The only crops that will thrive under extremely wet circumstances that prevail in equatorial swamps are most aroid crops and the sagopalm. But aroid crops for top yields appear to need somewhat lower temperatures. So, the only crop of this group really adapted to the hot humid tropical swamp environment is the sagopalm; actually, this palm finds its natural habitat under such conditions.

Table 3. Comparison of Productivity of the Main
Moisture-rich Starchy Staples

Crop (Reference)	Reported High Yield per ha	High Productivity 10^3 Kcal x ha^{-1} x day^{-1} of vegetation	Average World Yield 1974*	Average Productivity 10^3 Kcal x ha^{-1} x day^{-1} of vegetation
Cassava (CIAT 1969)	100 t in 305 days	416	9.2 t in 330 days	35
Sweet potato (IITA 1976)	43.1 t in 122 days	354	9.2 t in 135 days	68
Taro (Plucknett et al. 1971)	128.7 t in 365 days	339	5.4 t in 120 days	43
Sagopalm (Flach 1977)	25 t dry starch in 365 days	275	-	-
Yam (Rehm et al. 1976)	60 t in 275 days	193	9.8 t in 280 days	31
Banana (Purseglove 1972)	75 t in 365 days	155	12.5 t in 365 days	26

* After FAO Production Year Book 1976

Growth Habit

All the crops considered usually are propagated vegetatively, some by means of stem parts or suckers, others a part of the produce harvested. This presents a disadvantage for the latter category.

In general the growth cycle of the crops can be distinguished into the following phases:

1) Establishment - The plant parts used for propagation establish themselves. For perennial plants like sagopalm and banana this only is of importance for the first crop planted.

2) Development - The established plant develops its full leaf area.

3) Starch accumulation - Usually after formation of leaf area the plants start forming their sink and the accumulation of starch.

4) Ripening - Not all crops show ripening: a diminishing of the leaf area accompanied by a slackening starch accumulation.

Sagopalm and plantain show this phase clearly in fruit formation. But sagopalm may be harvested before this period starts.

A comparison of the phases and their duration is given in
Table 4 (v. also Flach 1980). Here both plantain and banana
clearly show some of the advantages of being permanent crops; but
plantain also shows the disadvantage of the produce being the
fruit.

During the phase of quick starch accumulation, partitioning
of dry matter over the storage organ and the other plant parts
appears to be constant (Boerboom 1978; Flach 1980), viz. a cons-

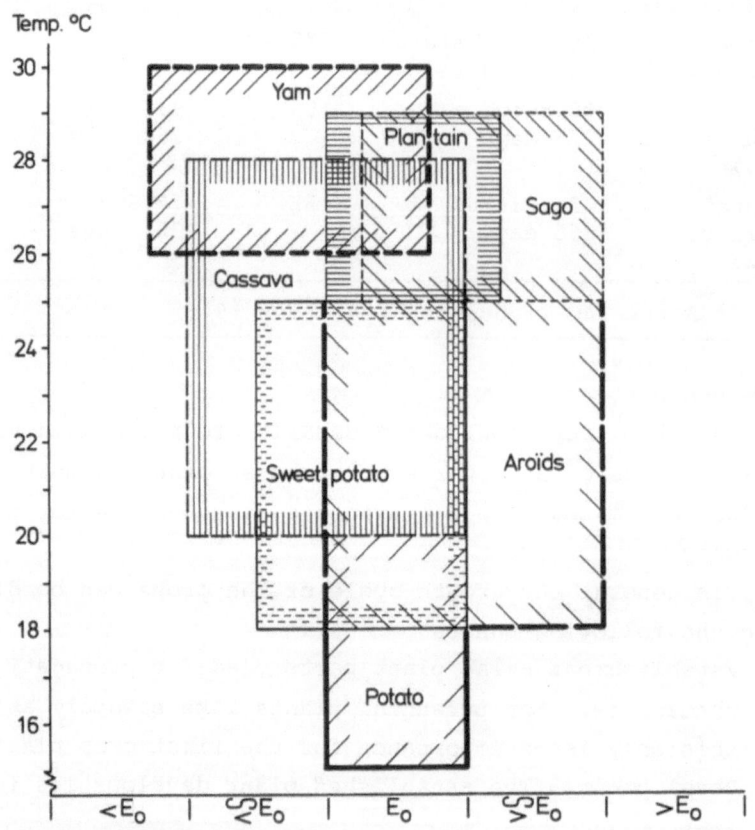

Figure 2. Combination of Water Requirements and Temperatures
for Optimal Growth of Several Staple Crops.
Temperatures are averages; water requirements are
expressed in the ability to tolerate water levels that
exceed or are less than potential evapotranspiration.

Table 4.   Growth Habit of Starch Crops

Crop	Total Duration of Growth	Soil Preparation	Establishment	Development	Quick Starch Accumulation	Ripening
	weeks	% of	total	duration	of	growth
Potato	15	13	13	20	55	–
Sweet potato	25	8	4	16	72	–
Yam	47	4	13	23	60	–
Aroids	34	6	9	15	70	–
Cassava	44	4	9	30	57	–
Sagopalm	390	–	–	23	77	–
Plantain	44	–	–	68	27	5

(after Flach 1980)

tant part of total dry matter production is used for starch, for leaves, etc. So the speed of leaf formation would be a good indicator of the speed of starch accumulation. This - not yet completely proved - general rule could be a good parameter for observations on the growth of the sagopalm.

THE SAGO PALM
Growth Habit

Most palms of the genus Metroxylon are hapaxantic (possess stems or trunks that only flower once) and soboliferous (give tillers or suckers that may grow horizontally initially and later on into erect stems). Such a plant thus develops into a cluster.

Most tillers or suckers eventually develop into a trunk. The trunk acts as a sink; in the pith of the trunk the superfluous starch produced by the leaves (the fronds) on top of the trunk is collected. Part of this starch, especially from the lower part of the trunk, may be used for suckering.

At the end of its life cycle each trunk produces an enormous

inflorescence on which a large part of the starch is expended.
After seed formation the trunk dies and deteriorates quickly.
Other suckers then, which usually already have developed into
separate trunks during the life cycle of the first trunk, take
over. Under favourable conditions each cluster of trunks may
produce one flowering trunk per two years or even two trunks
per three years.

After flower initiation the leaves diminish in size (Kiew
1977; below); this is accompanied by a diminishing starch
production. Flach (1977) gave an estimate of starch production
in relation to both age and number of leaves formed. During the
period of development two leaves per month may be formed. During
rapid starch accumulation only one leaf per month is formed, as
is normal in most palms (v. also Oijen 1909; Kiew 1977). The
basic system of starch accumulation is presented in Figure 3
(which is a corrected copy of Figure 4 from Flach 1977).

Planting of well sized suckers probably will show a clear
advantage over planting of seedlings. The period of development

photo: KT

A Sagopalm at Incipient Flowering (left rear), displaying
reduced frond growth compared with neighbouring palms, in
Singapore Botanic Gardens, mid-1979.

will be shortened. This may be even more true for suckers
attached to the mother cluster; they may be able to derive starch
from other trunks from the mother cluster.

At flower initiation the starch accumulation curve flattens
off again. If one aims at high production of starch per unit
area and time, trunks should be harvested just before flower
initiation. This actually is done in cultivation. But in wild
stands, where trunks may be abundant, the collector looks for
the highest production per trunk; this moment is around fruit
formation, probably just before.

Only under Sarawak conditions studies on starch content and
age have been undertaken (SAR 1970, 1973, 1974); Sim and Ahmed
(1977) report on these. Some earlier data from Sarawak were
used by Flach (1971); they are presented in Figure 4. A sample
of 95 sagologs each about 90 cm (3 ft) long from a factory in
Sarawak gave a result of 15-25% starch on wet basis, with an
average of 19%.

From the studies of van Gorkom (1956) and Colon (1958) the
following data were derived. The "average" trunk from New Guinea
consists of:

trunk weight	1250 kg	100%	—
cortex	400 kg	32%	—
pith	850 kg	68%	100%
starch	250 kg	20%	29%
water	425 kg	34%	50%
remainder	175 kg	14%	21%

The cortex, the outer 2-3 cm of the trunk, contains about 2.5%
starch on wet basis; in this trunk that would have amounted
to 10 kg of starch.

Natural Stands

Probably the true sagopalms of the genus Metroxylon at the
moment are the most underexploited plants on our earth. These
palms occur in large natural stands in equatorial swamps in
Southeast Asia. In West New Guinea, the Indonesian province of
Irian Barat, a survey of such an area was made. Between 1°44'

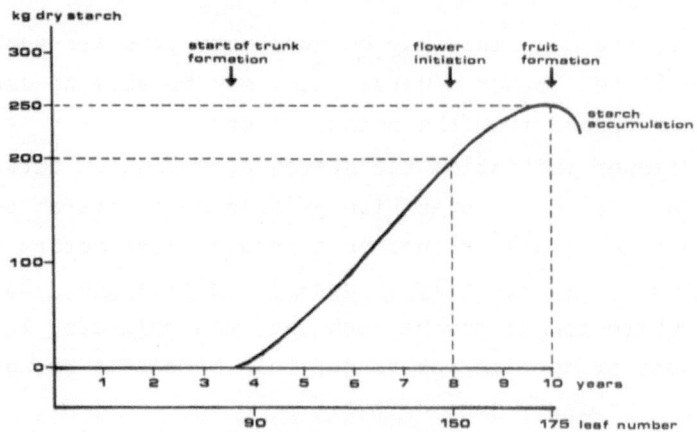

Figure 3. Assumed Rate of Starch Accumulation in a Sagopalm Trunk on a Clayish Soil.
The accumulation is presented as a function of time and of leaf number, and the physiological stages are given.

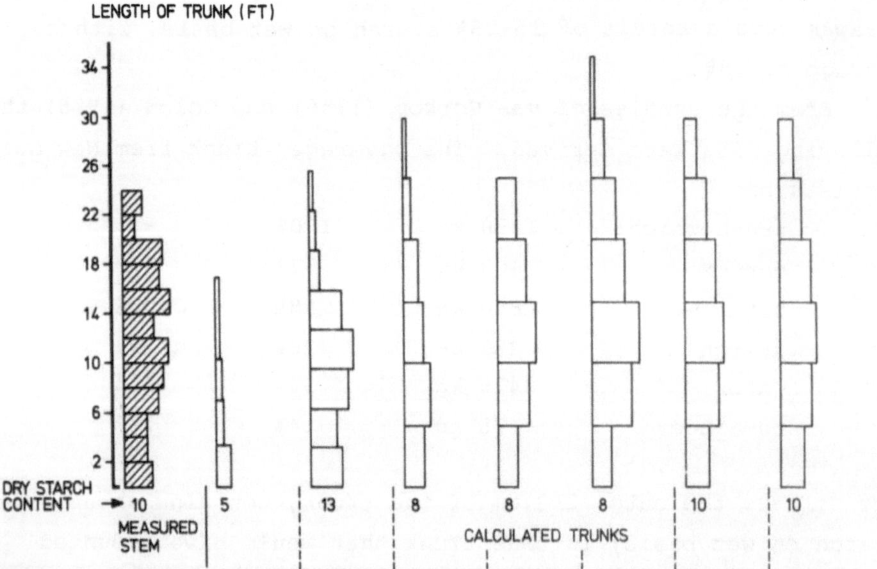

LENGTH OF TRUNK (FT)

DRY STARCH CONTENT →	10	5	13	8	8	8	10	10
	MEASURED STEM				CALCULATED TRUNKS			
WEIGHT OF TRUNK (LBS)	2000	1300	2100	2250	2000	2500	2250	2250
LENGTH OF STEM (FT)	24	17	25	30	25	35	30	30
TOTAL DRY STARCH YIELD (LBS)	320	25	250	175	280	390	325	325
STAGE OF DEVELOPMENT	JUST FLOWERING	YOUNG	YOUNG	JUST BEFORE FLOWERING	FLOWERING	FLOWERS AND YOUNG FRUITS	YOUNG FRUITS	WELL DEVELOPED FRUITS
ESTIMATED AGE (YEARS)	10	4	8	10	10.5	11	12	13

Figure 4. Starch Yield, Stage of Development, Size and Weight of a Number of Sagopalms in Sarawak

to $2°16'S$ and $132°$ to $133°30'E$, Stellingwerf (1957) investigated such an area from the interpretation of aerial photographs; it was estimated to contain at least 100,000 ha in which sagopalm dominated.

Earlier, in a ground survey, Zwollo (1950) estimated the same area to possess a production capacity of on the average 2.5 tons of dry starch per ha and year. His estimate was deliberately low because of (i) the extremely crude methods of starch extraction used by the local population, and (ii) the lack of knowledge on the age of the trunks harvested. But his estimate already represents over 27,000 Kcal. $ha^{-1}.day^{-1}$, and thus is of the same level as plantain in Table 3.

In the area in 1950 there were living some 15,000 inhabitants needing roughly 10% of the estimated production capacity for food; the remainder was going to waste. It is certain that the area under sago-dominated swamps on and around the island of New Guinea is several hundred thousand hectares. Wttewaal (1954) counted the number of clusters of sago, ripe trunks and overripe trunks on some five representative quarters of hectares in the same area; his results are presented in Table 5. Unfortunately his views on "ripe" and "overripe" were not specified in the report; it probably is the number of flowering trunks before and after seed formation. Also it is not stated whether trunks had been harvested in the areas measured.

Table 5. Composition of Natural Stands of Sagopalms

Observation	Old Trunks	Ripe Trunks	Clusters
1	30	20	160
2	-	33	200
3	22	33	245
4	-	-	220
5	20	33	260
Average	14.5	24	217

(after Wttewaal 1954)

Nevertheless, some conclusions can be drawn. Each cluster on the average occupies 50 sq m; this leads to an average distance

Table 6. Production Capacity of Sagopalms

Situation	Distance between clusters m	Mature Trunks no. per ha	Recoverable Starch kg per trunk	Possible Production of Dry Starch kg.ha⁻¹.yr⁻¹	Energy Production 10⁶ Kcal.ha⁻¹.yr⁻¹
As estimated by Zwollo (1950), Wttewaal (1954)	7	24	120	2,880	11.5
At improved starch extraction	7	24	185	4,440	17.8
In improved natural stands	7	55	185	10,175	40.7
In fully improved natural stands	7	136	185	25,160	100.6
At full cultivation	6	138	185	25,530	102.1

Table 7. Properties of Varieties of Sagopalm

Local Name	Height	Girth	Spines no.	Spines size	Leaves V-shape	Leaves Fit for atap(a)	Rachis Fit for gabah₂(b)	Leaf Sheaths Remaining on trunk	Starch Colour
Sagu mulat	4	1	1	-	no	1	2	3	white
Sagu tuni	1	2	1	2	no	2	1	2	white
Sagu ihur	1	3	2	3	no	2	1	1	reddish
Sagu makanaro	3	1	1	1	yes	2	2	3	white
Sagu ikau	2	3	2	1	no	3	2	1	reddish

(a) atap: roofing (b) gabah-gabah: rachis used for walls

1-4 : tall; large; many; excellent quality; continuously
    - small; none; worst quality; not at all

between clusters of 7 m. This is very close to Flach's (1977) observation on cultivated sago in Batu Pahat. There, planting distances were used of 6 x 6 m resulting in 277 clusters per ha and 7 x 7 m resulting in 205 clusters per ha. Such plantings gave high yields, if carefully tended. From 277 clusters per ha one trunk per cluster per two years were harvested, or 138 trunks per year; from 205 clusters per ha two trunks per three years were harvested or 136 trunks per year.

The area was estimated by both Zwollo (1950) and Wttewaal (1954) to contain 30-50% other trees. It stands to reason that starch production of such an area could be increased considerably by means of the following measures:
1) Poisoning or cutting of all other trees ;
2) Regulating of trunk growth through pruning of suckers ;
3) Earlier harvesting of ripe trunks, preferably before flower initiation ;
4) Careful methods of starch extraction.
An estimate of the possibilities is given in Table 6. These calculated yield possibilities of course need to be proven in practice. But it appears very reasonable that introduction of methods of cultivation will lead to at least a fourfold starch production.

The high production potential of natural stands of sago has been noticed often before. Salverda (1938; 1939) gives a good example. Production at a level of 10 tonnes per year from the 100,000 ha mentioned would give 1,000,000 tonnes or 4×10^{12} Kcal of dry starch per year. This would represent already 1% of world energy production of the moisture-rich starchy staples (Table 2).

Ecological Conditions

The sagopalm is found between 10°N and S up to an altitude of about 700 m. Information on ecological conditions for good sagopalm growth is rather scarce. One has to rely on natural habitat and some additional information from the few centres of cultivation. The limiting factor probably is the temperature, which outside this region occasionally may drop below 15°C.

Cultivation of sagopalm in West Malaysia probably is only possible on a sustained yield basis because of the nutrient content of the water that regularly floods the soil (Flach 1977). Preferably all material except starch should remain in the planting in order not to deplete the soil of nutrients.

Wee (1977) summarised the visual quality assessment of sagologs delivered to starch factories in Sarawak. An explanation of the points she mentioned is offered by me. A good quality sagolog:

a) will be submerged by at least 75% in water. The specific gravity of the moisture-rich starchy staples usually correlates linearly with their starch content. This was shown to be true also for sagologs (SAR 1975: 219). Both girth and fresh weight of logs were found to be linearly correlated with the starch yield.

b) possesses bulging midsection. After cutting of a trunk from a cluster the next largest trunk suddenly is exposed to full sunlight; this probably causes the trunk to enlarge its girth.

c) has narrow spacing between the leaf scars. This points to a partitioning of dry matter favourable for starch production. Leaf production probably is regular at the rate of one leaf per month and partitioning of dry matter is constant.

d) possesses coarse, thick bark. This feature appears to be related to somewhat brackish water. It was already shown by Flach et al. (1977) that sagopalm seedlings were not harmed by salinity levels of up to EC = 4.5 mmhos. There even is some reason to assume that salinity might be advantageous. This confirms the opinion of sago growers in both Sarawak and West Malaysia. Moreover, Stellingwerf (1957) shows that sagopalm swamps often border on mangrove and nipah swamps.

Research on growth and yield of starch under varying soil and light conditions in Sarawak (SAR 1973) revealed that sago trunks grown on acid peat were somewhat less in girth and weight than trunks from mineral soils; also the starch content on peat soils was less. It appears that both growth and starch accumulation are retarded on peat soils. This probably is due more to low fertility of peat soils than to low pH. In its natural habitat the sagopalm is not found on peat soils, although the soils may contain appreciable organic matter (up to 20%).

In general we now may conclude that good ecological conditions for the sagopalm are:
1) a minimum temperature of $15^{\circ}C$;
2) full sunshine;
3) regularly but not continuously flooded situation;
4) floodwater with an appreciable nutrient content and possibly slightly brackish;
5) mineral, preferably clayish soils with organic matter content of up to 20%;
6) pH 4 and higher.

Varieties

Evidence on the validity of distinguishing a number of species of the true sagopalm in the natural stands on and around New Guinea still is controversial. But it is very clear that the local population distinguishes a number of varieties.

Sastrapradja and Mogea (1977) gave five local names from Ambon and mentioned their spine characters. The same local names were given by Tupamahu (1909) for Ambon, by Fortgens (1909) for Halmahera; Zwollo (1954) made approximately the same distinctions in New Guinea, so did Salverda (1938) on Celebes (Sulawesi). Table 7 is an attempt to combine all qualities of trunks per variety distinguished by local people, prepared in cooperation with an informant from Ambon (Reawaru 1980; pers. comm.). The locally distinguished varieties still are in accordance with Rumphius (1741).

It remains, however, uncertain whether and to what extent we have to do with species. It is for instance reasonably certain that spines as a distinguishing character in itself would be insufficient. Further research into this matter is much needed. It may well be that different species differ somewhat in growth pattern and starch yield. Moreover, the spines make access to natural stands more difficult, even if they are a protection against animal attack.

REALISATION OF THE POTENTIAL OF NATURAL STANDS

It is amazing that such a potential production, high even when compared to modern methods of cultivation of other crops, has not been obtained, especially if one realises that the region possesses several hundred thousand hectares of sagopalm stands. The main reason for this appears to be that the wild stands usually are highly inaccessible. It is a swamp with a clayish soil, which means that already the movements of man are rather limited. It is moreover jungle: a complicated, dense mass of fallen and rotting leaves, branches and trunks, in the case of sagopalm with the accompanying spines. Special measures, requiring imagination, are necessary to open up such an area and to organise production.

The average sago trunk weighs some 1000 kg; up to 80% of this trunk may consist of water. This is the main reason why the local inhabitants, experienced swamp walkers as they are, do not attempt to remove sagologs from the swamp. They fell trunks and process them in the swamp; only the wet starch is taken home. Harvesting from natural stands therefore would have to be organised in a comparable way, as Salverda (1938) already showed. It would probably be best to locate a factory for starch, feed or alcohol in a convenient place within or in the neighbourhood of such a swamp. This factory could then use the raw material from the swamp.

A possibility might be to work with a number of lighters. On such a lighter water cleaning equipment might be necessary. From the lighter clear water is pumped through a movable pipeline to the place where a number of people fell sagopalms. There a mill grinds sago pith, the ground pith with the clear water being washed back to the lighter through a second pipeline. The mill is mounted on a sled.

The workers are equipped with chain-saws, axes and sleds for transport of the pith. When a tree is felled its fronds are cut off and spread over the soil. The trunk is trimmed, split into halves parallel to its length. The pith is dug out, put into the sleds and brought to the mill. At harvesting twice a year also the other operations are done, like pruning of the clusters of palms, weeding of all other trees and possibly replanting of unsuitable material.

Some systematic research in this field has been done by a Dutch company in the period 1957-1962. This company probably possesses a number of specifications on how such an operation could be run in practice.

POTENTIAL

The underexploited true sagopalms of the genus Metroxylon at present contribute only 1.5% to the total energy produced by the main moisture-rich starchy staples in the tropics. The palm could easily play a prominent role if the 0.5 million ha of natural stands on and around the island of New Guinea were developed.

Despite a general lack of detailed agronomic knowledge on the crop, a yield of 10 tonnes of dry starch per ha and year may be expected, which under further improved conditions could rise to 25 tonnes. Improvement of the natural stands in equatorial swamps should be combined with harvesting and processing, for which very special measures will be needed. These measures are possible at today's level of technology and prospects of using biomass production.

It appears that, due to mainly the economic climate, the until now completely neglected starch producing crop, the sago-palm, will obtain its rightful place among the tropical moisture-rich starchy staples. If its starch from natural stands could be produced sufficiently cheaply it could well be a renewable and thus continuous source of biomass energy.

It is interesting to note that the Japanese have similar ideas, considering the special issue of the Japanese Journal of Tropical Agriculture, volume 23 (1979: 117-171), with ten papers on the sagopalm.

REFERENCES

BOERBOOM, B.W.J. 1978. A model of dry matter distribution in cassava (Manihot esculanta CRANTZ), Neth. J. agric. Sci. 26: 267-77.

CIAT 1969. Annual Report of the Centro International d'Agricultura Tropical; Cali (Colombia).

DE VRIES, C.A., J.D. Ferwerda and M. Flach 1967. Choice of food crops in relation to actual and potential production in the tropics, Neth. J. agric. Sci. 15: 241-8.

FAO 1976. FAO Production Year Book; Rome.

FLACH, M., A.H.J. Kroon and J.S. Baker 1971. Sago Production in Pahang Tenggara; Pahang Tenggara Regional Masterplanning Study, pap. no. 16.

FLACH, M. 1977. Yield potential of the sagopalm and its realisation, op. cit., ed. TAN: 157-77.

FLACH, M. 1980 (in press). Ecological competition among the main moisture rich starchy staples in the tropics and subtropics; Proc. 5th Inter. Tropical Root Crops Symp., Manila, 1979.

FLACH, M., F.J.G. Cnoops and G.C. van Roekel-Jansen 1977. Tolerance to salinity and flooding of sagopalm seedlings: Preliminary investigations, op. cit., ed. TAN: 190-5.

FORTGENS, J. 1909. De sagoepalm, Koloniaal Mus. te Haarlem Bull. no. 44: 71-104.

I.I.T.A. 1976. Annual Report of the International Institute for Tropical Agriculture; Ibadan.

KIEW, R. 1977. Taxonomy, ecology and biology of sago palms in Malaya and Sarawak, op. cit., ed. TAN: 146-54.

OIJEN, L.A.T.J.F. 1909. Sagoe en sagoepalmen, Koloniaal Mus. te Haarlem Bull. no. 44: 12-69.

PAHANG TENGGARA 1972. The sago experimental project; Pahang Tenggara Regional Masterplanning Study, appendix 4: 34.

PLUCKNETT, D.L., and R.S. de La Pena 1971. Taro production in Hawaii, World Crops 23: 244-9.

PURSEGLOVE, J.W. 1974. Dicotyledons; Longmans, London: 719.

REHM, S., and G. Espig 1976. Die Kulturpflanzen der Tropen und Subtropen; Ulmer Verlag, Stuttgart: 321.

RUMPHIUS, G.E. 1741. Het Amboinsche Kruyd-boek; Burmanus, Leiden: 72-83.

SASTRAPRADJA, S., and J.P. Mogea 1977. Present uses and future development of Metroxylon sagu in Indonesia, op. cit., ed. TAN: 112-7.

SAR 1970, 1973, 1974, 1975. Annual Reports of the Research Branch of the Department of Agriculture Sarawak; Gov. Printer, Kuching (Chapters "Chemistry").

SALVERDA, Z. 1939. Ervaringen van een landhuishoudkundig onderzoek op Nieuw Guinea (Experiences in an economic exploration of New Guinea), Bergcultures 13: 1265-76.

SIM, E.S., and M.J. Ahmed 1977. Variations of flour yields in Sarawak sagopalms, op. cit., ed. TAN: 178-80.

TAN, K. (ed.) 1977. Sago-76: Papers of the First International Sago Symposium; Kuala Lumpur.

TUPAMAHU, J. 1909. Pokok sagoe atau pokok roembia (The sago or rumbia tree), Koloniaal Mus. te Haarlem Bull. no. 44: 105-12.

WEE, A.C. 1977. Discussion, op. cit., ed. TAN: 288.

Depositions in the Archives of the former Directorate Netherlands New Guinea:

COLON, F.J. 1958. Het winnen van sago in de moerassen van Nederlands Nieuw Guinea (Production of sago starch from the swamps in Netherlands New Guinea); TNO Report no. 58-0705, The Hague.

SALVERDA, Z. 1938. Groot-exploitatie van sago (Large scale sago exploitation).

STELLINGWERF, D.A. 1957. Rapport bij de Vegetatieschetskaart van Amaroe-Zuid, Inanwatan en Sebjar-Babo, gelegen in de Vogelkop Nieuw Guinea (Report with the provisional vegetation map of Amaroe-South, Inanwatan and Sebjar-Babo in Cenderwasih, Irian Barat); I.T.C. report, Delft.

VAN GORKOM, C. 1956. Het winnen van sago in de moerassen van Nederlands Nieuw Guinea (Production of sago starch from swamps in Netherlands New Guinea); TNO Report no. 232, The Hague.

WTTEWAAL, B.W.G. 1954. Rapport betreffende de mogelijkheden van de oprichting van een mechanisch sagobedrijf te Tarof (Report on the possibilities for a mechanised sago operation at Tarof).

ZWOLLO, M. 1950. Rapport sago onderzoek Inanwatan (Report on investigations into sago production at Inanwatan).

PHYSIOLOGY OF BIOMASS PRODUCTION IN THE TROPICS

M.M. LUDLOW

PHOTOSYNTHETIC SYSTEMS

Photosynthesis is a biological process which converts renewable solar energy into chemical energy in the form of biomass.
The efficiency with which this conversion occurs is determined
by the seasonal input of solar energy, water and carbon dioxide,
and can be modified by high- or low-temperature stress or by
availability of soil nutrients. The agronomist and plant breeder
can attempt to control such photosynthetic systems, both by
modifying inputs or constraints, or by improving the efficiency
with which the crop converts these inputs. Because it is technically, if not economically, feasible to control the supply of
water and mineral nutrients, and because temperature stress is
less in the tropics than in other climatic zones of the world,
I will concentrate on the efficiency with which plants convert
solar energy received into biomass.

In the absence of climatic and soil constraints, biomass
production depends upon the amount of solar energy received,
extent of and manner in which it is intercepted, and the photosynthetic and respiratory characteristics of the plant. Yield
of a particular part of the biomass (economic yield) such as
grain or storage product, e.g. starch in sagopalm, depends upon
how much of the total biomass production is partitioned into
that product.

There is virtually nothing known about the production
physiology of sago, and little reliable yield data are available.
Therefore it might be useful to present the general principles
of biomass production and indicate how they might be applied
to sago production.

W.R. Stanton and M. Flach (eds.), SAGO. The Equatorial Swamp as a Natural
Resource. Proceedings of the Second International Sago Symposium. All rights reserved.
Copyright ©1980 Martinus Nijhoff Publishers, The Hague/Boston/London.

MAXIMUM GROWTH RATES AND NET ANNUAL PRODUCTION:
POTENTIAL AND ACTUAL

Where climatic and soil stresses are minimal, the upper limit
of biomass production is set by the amount of solar radiation
received. Solar radiation arriving at the earth's surface in
the Wet Tropics varies with latitude and season between 11 and
22 MJ m^{-2} d^{-1}, the average being ca. 17.2 MJ m^{-2} d^{-1} (Cooper 1975).
The potential maximum crop growth rate for an area receiving
this average figure was calculated using the method of Loomis
and Williams (1963). The calculation and the assumptions made
are given in Table 1, and the relationship between crop growth
rate and daily solar radiation shown in Figure 1. A figure of
63 g m^{-2} d^{-1} was obtained which is equivalent to an annual net
dry matter production of 230 t ha^{-1}; these represent an efficiency
of conversion of solar energy of 6.5%[a].

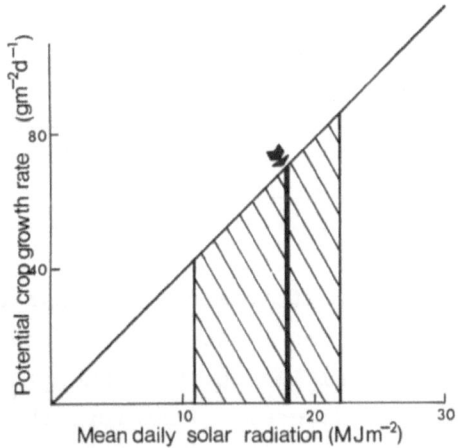

Figure 1. Relationship between Mean Daily Solar Radiation and
          Potential Crop Growth Rates, also showing range of
          values recorded in the Wet Tropics (shaded area) and
          mean value used in calculations (17.2 MJ m^{-2} d^{-1}).
          (Calculated by the method of Loomis and Williams 1963)

[a] Many authors quote conversion efficiencies as a percentage
of photosynthetically-active radiation (PAR), but I intend to use
solar (or short-wave) radiation because this conforms with meteo-
rological records and no assumptions are necessary about the proportion which is photosynthetically-active. PAR is approximately
50% of solar radiation in the Tropics (Monteith 1972), so efficiency on that basis is twice those quoted here.

Table 1.   Potential Production in the Wet Tropics

Growth or Production Factor	Value
Average daily solar radiation	$= 17.2$ MJ m^{-2}
Total quanta, 400-700 nm, (2.063 μE J^{-1} solar radiation)	$= 35.48$ E m^{-2}
Albedo loss ($\sim$8.3%)	$= -2.94$ E m^{-2}
Inactive absorption ($\sim$10%)	$= -3.55$ E m^{-2}
Total quanta, 400-700 nm, available for photosynthesis	$= 29.0$ E m^{-2}
Amount of carbohydrate (CH_2O) produced (quantum requirement $= 10$)	$= 2.9$ mole m^{-2}
Respiration loss ($\sim$33%)	$= -0.96$ mole m^{-2}
Net production of carbohydrate (CH_2O)	$= 1.94$ mole m^{-2}
Net production (30 g mole^{-1} $\{CH_2O\}$)	$= 58$ g m^{-2}
Net dry matter production (DM = 92% $\{CH_2O\}$)	$= 63$ g m^{-2}
Annual dry matter production (365 days at 17.2 MJ m^{-2} d^{-1})	$= 230$ t ha^{-1}
Efficiency of conversion of solar radiation (1 g dry matter = 17.8 kJ)	$= 6.5$%

Note: Assumptions used in the calculations are given in
       parenthesis.

(after Loomis and Williams 1963)

Leaves only achieve quantum efficiencies as high as those
used in the calculation under light-limited conditions. Due to
variations in display, individual leaves in a canopy may be
exposed to a wide range of irradiances. Therefore the curvili-
nearity of the light response curve of photosynthesis and the
non-uniform illumination must be considered to obtain a more
realistic estimate of maximum production for specific crops
and environments.

Simulation models have been used to overcome these problems;
they also allow photosynthetic characteristics for individual
crops to be included. Duncan's model gives figures of 50-80
g m^{-2} d^{-1} for maize at Davis, California, which is equivalent
to an efficiency of conversion of 3% for the higher solar radiation they receive during summer (Loomis et al. 1971). The
potential calculated for this situation using Loomis and William's

method (Table 1) is 100 g m^{-2} d^{-1}. Using a similar modelling approach de Wit (1965) calculated a mean annual growth rate of 25 g m^{-1} d^{-1} or 91 t ha^{-1} over a year for tropical C$_4$ grasses in Uganda.

Monteith (1972) employs another approach to estimate maximum growth rates and yields for the Tropics. He multiplies solar constant by a series of efficiency factors which relate to: (a) geometry of the earth relative to the sun; (b) atmospheric transmission of solar radiation; (c) spectral changes; (d) photochemical process of CO_2 fixation; (e) diffusion process of CO_2 uptake; (f) interception of solar radiation; and (g) respiration. His estimates for annual production in the Tropics for intensive and subsistence agriculture, respectively, are 110 and 9 t ha^{-1}, with the maximum rates of dry matter production of crops intercepting 95% of average incident radiation at 34 and 18 g m^{-2} d^{-1} for C$_4$ and C$_3$ crops respectively; these figures represent efficiencies of conversion of 2.7% and 1.4% of solar energy received. Thus estimates of both maximum growth rate and annual yield from simulation studies which incorporate some of the inherent inefficiencies of plants are lower than potential yields in Table 1. However, the extent of the discrepancy varies with the model used and values of parameters for particular species and environments.

Maximum crop growth rates and net annual production (dry matter yield) actually recorded for a number of plants in the Tropics are given in Tables 2 and 3. The highest crop growth rate recorded of 54 g m^{-2} d^{-1} for Pennisetum typhoides approaches the potential value of 63 g m^{-2} d^{-1} and values calculated from Duncan's model; however, it exceeds Monteith's estimate of 34 g m^{-2} d^{-1}. Conversion efficiencies of these fast growing C$_4$ grasses approach but do not reach the 5-6% potential value (Loomis and Gerakis 1975), but the corresponding values for C$_3$ species are about half the potential. The higher crop growth rate of C$_4$ compared with C$_3$ species (Table 2) agrees with comparisons made on a wider range of species and climates (Stewart 1970; Loomis et al. 1971; Cooper 1975; Loomis and Gerakis 1975; Monteith 1978; the factor of two difference in crop growth rate

Table 2. Maximum Crop Growth Rates and Conversion Efficiencies of Solar Radiation in the Tropics

Crop	Location	Crop growth rate $g\ m^{-2}\ d^{-1}$	Solar radiation $MJ\ m^{-2}\ d^{-1}$	Conversion efficiency %
C_4 species:				
Pennisetum typhoides	N.T., Australia, 14°S	54	21	4.7
Pennisetum purpureum	El Salvador, 14°N	39	17	4.6
Sugarcane	Hawaii, 21°N	37	17	4.2
Maize	Thailand, 15°N	31	21	3.0
C_3 species:				
Rice	Philippines, 15°N	27	17	3.2
Cassava	Malaysia, 3°N	18	17	2.2
Stylosanthes humilis	N.T., Australia, 14°S	13	-	-
Oil palm (Elaeis guineensis)	Malaysia, 3°N	11	16	1.6

Source: Stewart 1970; Cooper 1975

between C_3 and C_4 plants may be fortuitous and depend upon the choice of species. Monteith (1978) found that the ratio for a wider range of species was 1.4, and the ratio for sugarcane and sugarbeet is 1.6 (Austin et al. 1978). Whatever the real figure is, it is certainly distinguishable and greater than 40%.

Even larger differences are found between mean crop growth rates over longer intervals, months rather than weeks (Stewart 1970; Monteith 1978), probably because C_4 species are mainly grown in the Tropics where radiation and temperatures are more favourable for plant growth than in higher latitudes where the majority of C_3 crops are grown, e.g. Monteith (1978) reports ratios of 1.7:1 for C_4:C_3 species. As a consequence, the net annual production of C_4 crop is greater than C_3 crop species, tropical rainforests, and rubber plantations (Table 3).

Yields of C_4 species vary between 16-85 t ha^{-1}, whereas yields of C_3 crops vary between 22-40 t ha^{-1}, a range which includes values for tropical rainforests and rubber plantations.

Table 3. Maximum Net Annual Production and Conversion Efficiencies of Solar Radiation in the Tropics

Crop	Location	Net production $t\ ha^{-1}$	Growing period day	Conversion efficiency Year %	Growth period
C_4 species:					
Pennisetum purpureum	El Salvador, $14°N$	85	365	2.7	2.7
Sugarcane (sugar)	Hawaii, $21°N$	64 (22)	365 –	2.0 (0.6)	2.0 (0.6)
Maize (grain)	Peru, $12°S$	26 (10)	–	0.8 (0.3)	–
Pennisetum typhoides	N.T., Australia, $14°S$	22	112	0.5	1.9
Maize (grain)	Thailand, $15°N$	16 (17)	103 –	0.5 (0.2)	1.3 (0.6)
C_3 species:					
Oilpalm (oil)	Malaysia, $3°N$	40 (5)	365	1.6 (0.4)	1.6 (0.4)
Cassava (Manihot esculenta) (tubers)	Malaysia, $3°N$	38 (22)	270 –	1.2 (0.7)	1.9 (0.9)
Leucaena leucocephala	Qld., Australia, $17.5°S$	32	365	–	–
Rice	Philippines, $15°N$	22 (12)	205	0.7 (0.4)	1.3 (0.6)
Mixed Cropping System:					
Rice and sorghum (grain)	Philippines, $15°N$	– (23)	365	– (0.8)	– (0.8)
Forests:					
Tropical rainforest	Africa	32	365	–	–
Rubber	Malaysia	35	365	1.4	–

Source: Stewart 1970; Corley et al. 1971; Loomis et al. 1971; Cooper 1975; Ferraris 1979.

The discrepancy between yields of C_4 and C_3 species increases with the length of the growing season (Figure 2). When crops or forests which grow for a full year are compared (Table 3), the average net production of C_4 species was about twice that of C_3 species; in addition the superiority of C_4 species increases

134

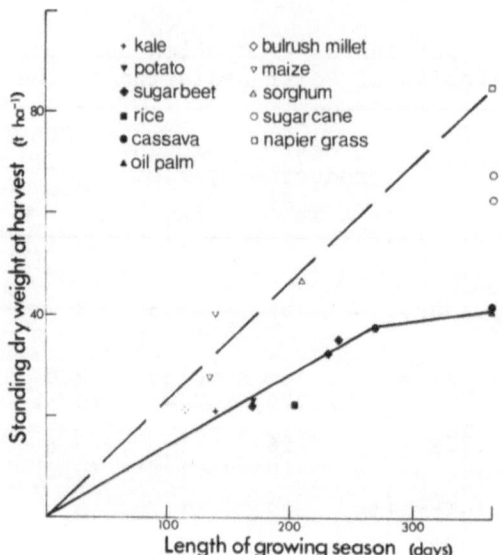

Figure 2. Standing Dry Weight at Harvest of C₃ and C₄ Crops
in Relation to Length of Growing Season.
C₃ and C₄ refer to the Calvin and C₄ dicarboylic
acid pathways of photosynthesis, respectively
(Monteith 1978).

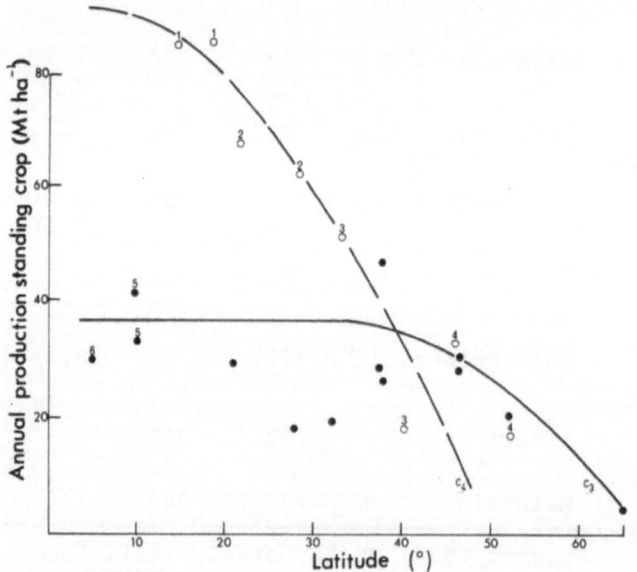

Figure 3. Annual Net Production of C₃ and C₄ Crop Species
in Relation to Latitude.
Numbers associated with points refer to species:
1) Pennisetum purpureum; 2) Sugarcane; 3) Sorghum spp.;
4) maize; 5) cassava; 6) oilpalm
(Loomis and Gerakis 1975).

from high to low latitudes, such that in the Tropics the net
annual production of C_4 and C_3 also differed by a factor of 2
(Figure 3).

On the other hand, yield of economic products such as grain,
tubers or sugar are rather similar among all crop species, with
a maximum value of 23 t ha^{-1} a^{-1} for a C_3 crop, a C_4 crop, and
a cropping sequence of C_3 and C_4 grain crops (Table 3). Efficiencies of conversion are clearly higher for the main growing
period than for a full year and for total biomass than for economic yield. Furthermore values for net annual production (Table
3) are less than for peak growth rates (Table 2). Efficiencies
of conversion for biomass production in C_4 species approach 3%,
whereas those for C_3 species approach 2%; the corresponding
figure for economic yield is 1% for both C_3 and C_4 species.

Even the maximum yield of 85 t ha^{-1} (Table 3) or 88 t ha^{-1}
(Stewart 1970) for elephant grass with an efficiency of conversion
of 2.7% falls below the potential of 230 t ha^{-1} and 6.5%.
Obviously yields and efficiencies of other species are further
below the potential. However, yields of elephant grass do
approach the more "realistic" estimates of 91 t ha^{-1} calculated
for the Tropics by simulation for a C_4 crop (de Wit 1965) and
the 110 t ha^{-1} for a C_3 crop (Monteith 1972). Both potential
yields and these simulated values are based on many assumptions.
Therefore it is not possible to state categorically what the
maximum yield might be.

Clearly the potential represents the limit but estimates
from models, e.g. 100 t ha^{-1}, may represent a more attainable
limit, against which agronomists and plant breeders can gauge
actual yields. However, there are some reports of dry matter
yields for sugarcane between 110-150 t ha^{-1} a^{-1} (Bull and Glasziou
1975), which suggest that estimates from models may be conservative. Whatever the attainable maximum yield is, only a few
C_4 grasses approach the most conservative estimates of net annual
production. Yields of the remaining C_4 species and all C_3 crop
and forest species are much lower. Possible reasons for the
discrepancy between actual and potential growth rates and yields
are discussed below.

REASONS FOR DISCREPANCIES BETWEEN ACTUAL AND POTENTIAL GROWTH RATES AND YIELDS

There is a multitude of reasons for farm yields being less than the potential. Yields are depressed by deficiencies of mineral nutrients and water, by effects of pathogen and insect attack, and by other soil characteristics. Tropical soils are inherently infertile, and water supply restricts yield in the monsoonal Tropics. All these are, to some extent, amenable to solution with our current level of agricultural technology, although which solutions are economical will vary with the crop and the location. There are other situations where economic, social, religious or cultural factors result in lower yields of a particular crop, e.g. where mixed cropping is practised, or where economic yield is sacrificed so that some other part of the plant can be used for fuel, shelter, clothing etc.; although they may be very important in some circumstance these factors will not be discussed because they are specific to certain countries, religions, races and ways of life. Instead I am going to concentrate on more general factors which will be common to all countries, and on the discrepancy between the highest actual yields (rather than farm yields) and potential yields.

Leaf Photosynthetic Characteristics

The difference in leaf photosynthetic characteristics between C_3 and C_4 plants is now well established (e.g. Hatch et al. 1971; Ludlow 1976). At temperatures above ca. 30 °C the quantum efficiencies (net photosynthetic rate per unit of light absorbed at low light levels) of C_4 plants are greater than those of C_3 plants, and the difference increases with temperature (Ehleringer and Bjorkman 1977); in addition, rates of leaf net photosynthesis of C_4 plants in full sunlight are approximately double those of C_3 plants. In some comparisons, this factor of two difference is propagated from leaf photosynthesis to growth rate and finally to the yield level (e.g. Ludlow and Wilson 1972). However, there are instances where the higher leaf photosynthetic rate of C_4 plants is offset by some deficiency, e.g. leaf area development, such that some C_4 plants are not superior in growth rate or yield

(v. Bull, and Slatyer, in Hatch et al. 1971). Thus possession of the C_4 pathway should be considered only as a potential advantage for higher growth rates and yields.

Notwithstanding these reservations, it is evident that many C_4 plants are superior to C_3 plants in both maximum and long-term growth rate (Table 2) and net annual production (Table 3; Figures 2 and 3); this superiority increases with length of growing season and from high to low latitudes. The extent of the advantage varies from ca. 2:1 for net annual production (Figure 3), to ca. 1.7:1 for long-term growth rate and efficiency of conversion (Cooper 1975), and to 1.4:1 for peak growth rates (Monteith 1978). Moreover peak growth rates of some C_4 plants approach the potential, whereas C_3 plants to only about half (Table 2). Thus based on measured values, lower growth rates and yields must at present be expected from C_3 plants (Cooper 1975; Loomis and Gerakis 1975).

Respiration

In the calculation of potential growth rates and yield, it was assumed that respiration was 33% of photosynthesis (Table 1). There may well be situations, times, and species where this figure underestimates the actual respiratory load, e.g. values of up to 50% have been recorded in a number of crop and pasture species (Monteith 1972), and even higher figures are found at some stages of development such as the later phases of grain filling in cereals. Also plants with a large biomass to maintain, such as trees, might be expected to have a higher respiratory load; indeed figures of about 75% have been recorded for tropical rainforests (Monteith 1972). In addition, the proportion of photosynthate used in respiration is probably larger in tropical than in temperate climates because the temperature coefficient for respiration of green tissue exceeds the coefficient for photosynthesis over the range of temperatures in which plants usually grow.

Thus in a plant such as sagopalm where the stem acts as a living biological storage container which has to be maintained for the life of the tree, the proportion of photosynthate used in respiration is probably greater than that for herbaceous crops and the assumed figure of 33%.

Temperature

Whereas temperatures in the Tropics are the most favourable for plant growth of all the climatic zones of the world, temperature still imposes a limitation to biomass production. The decline of temperature and the consequent limitation imposed on plant growth with increasing altitude in the Tropics is well known. Also the seasonal change and its effect of plant growth increases with latitude (Figure 3) (Loomis and Gerakis (1975) assumed the relationship between yield of C_4 plants and latitude was due to difference in solar radiation. I believe that temperature is a more important factor); e.g. biomass production of both C_3 (Leucaena leucocephala) and C_4 (Pennisetum purpureum) plants at South Johnstone, Australia (17°S) shows a marked seasonal variation, which is due mainly to variation in temperature (Ferraris 1979; Ferraris and Stewart 1979). Pasture growth in northern tropical Australia can also be limited by temperature in winter, if soil moisture is adequate.

The optimum temperatures for photosynthesis and growth of C_4 plants are higher than those of C_3 plants (Ludlow 1976). However, I am not aware of any reports of biomass production of C_3 plants being restricted by supra-optimal temperatures in the Tropics.

Light Interception

I believe the amount of and manner in which solar radiation is intercepted to be the major factor associated with the depression of yield below the potential (v. discussions by Loomis and Williams 1963; Monteith 1972). On an annual basis the amount of solar radiation absorbed by green tissue is directly related to the length of time a full vegetative cover is maintained; Monteith (1978) shows how the yield difference between and within C_3 and C_4 crops increases with the length of the growing season. The more favourable conditions for growth of both natural vegetation and agricultural plants in the Tropics is conducive to the evergreen habit and high light interception throughout the year,

compared with Temperate areas. Thus the opportunity for maximum
light interception exists in the Tropics, especially for
perennials.

Low light interception occurs during the establishment of
crops and forests; this period varies from a few weeks or months
in crop species to several years in plantation crops and forests
(Loomis and Williams 1963; Cooper 1975). Until full light interception is achieved there is a simple linear relationship between
solar radiation intercepted and production rates of both C_3 and
C_4 species (Loomis and Gerakis 1975). Obviously the longer the
period between sowing and full light interception, the lower will
actual yields fall below the potential. Other interception losses
can occur because of: (a) defoliation of pastures, (b) reduced
leaf area as a result of activities of insects, pathogens and
man, and (c) suboptimal planting density and arrangement or by
thinning.

As leaf area per unit ground area (leaf area index, L)
increases crop growth rate increases until full light interception
is achieved (Brown and Blaser 1968; Loomis and Gerakis 1975);
usually, once L increases above this value, there is no further
increase in growth rate over a wide range of L, i.e. a plateau-
shaped response. Thus no further increase in growth rate results
from additional investment in leaf area. However, in some species
and in some instances, crop growth rate declines when L exceeds
an optimum, i.e. an optimum-shaped response; in this case not
only is the extra investment in leaf area wasteful, but also it
reduces growth rate and yield.

Once full light interception is achieved, photosynthesis of
a particular canopy is determined by the manner in which the
light is intercepted, and the efficiency with which this occurs
determines how close actual yields approach the potential.
Theoretically, the canopy should be composed of leaves in such
a position so that the whole photosynthetic surface is working
at the maximum photochemical efficiency which is only achieved
at low light levels. This is approached by canopies with erect
leaves at the top and with leaf angle increasing towards the
horizontal with depth. Some canopies show this pattern but many

do not, especially those with horizontal leaf arrangement, in which case light levels on leaves are higher than desirable, particularly at the top of the canopy, and hence conversion efficiency and yield potential will be reduced.

PALM SAGO YIELDS AND SUGGESTIONS FOR IMPROVEMENT

Yields of sago starch quoted by Ruddle et al. (1978) varied between 70 and 500 kg per palm, with the higher yields coming from sterile palms. It would be interesting to know if the higher yield arose because these palms grew for a longer time or because sterility had a beneficial effect on yield; if it were the latter, the possibility exists of increasing yield by breeding, selecting or chemically inducing sterility. Because of the lack of accompanying data, great care is needed in calculating or interpreting annual production of sago starch. On the basis of their figure of 330 palms/ha harvested per year, if the starch yield were 100 kg per palm, yield of starch would be 33 t ha^{-1}, or 22 t ha^{-1} on a dry matter basis. This seems a very high number of palms harvested per hectare per year because the next highest is 30-60.

The highest yield recorded based on reasonably reliable data appears to be 60 palms harvested per hectare in year, with 110-136 kg of starch per palm, which is equivalent to 7-9 t of starch or 5 t on a dry weight basis (Purseglove 1972; Duke 1977). If we assume that the mean solar radiation was 17.2 MJ m^{-2} d^{-1}, this represents a conversion efficiency of 0.13%. The yield of 5 t ha^{-1} a^{-1} is much less than the 22 t ha^{-1} a^{-1} for cassava, and the conversion efficiency is considerably lower than values (range 0.2-0.9%) for economic yield of other C_3 and C_4 crops grown in the Tropics (Table 3). [b] However, most sago yields are much lower than this, e.g. Barrau (1959) reports dense groves of Metroxylon with very low amounts of starch per palm.

---

[b] Whereas efficiencies of conversion of solar energy into starch by sagopalms are low compared with other species in the Tropics, the efficiency expressed on the basis of labour, inputs or capital may be considerably more attractive. Such comparisons are outside the scope of this paper.

Table 4. Potential Yield of Sagopalm

		Number of palms harvested per annum					
		5	10	20	40	60	80
Starch yield kg per palm	50	0.2	0.5	1.0	2.0	3.0	4.0
	100	0.5	1.0	2.0	4.0	6.0	8.0
	200	1.0	2.0	4.0	8.0	12.0	16.0
	300	1.5	3.0	6.0	12.0	18.0	24.0
	400	2.0	4.0	8.0	16.0	24.0	32.0
	500	2.5	5.0	10.0	20.0	30.0	40.0

Sago starch t ha^{-1} a^{-1}

Table 4 shows the potential annual yield of sago (air dried) per hectare, calculated using recorded the range of sago yield per palm and numbers of palms harvested; the combinations of these two factors which gave the maximum recorded yield of 8 t ha^{-1} a^{-1} ($\equiv$ 5 t ha^{-1} a^{-1} on dry weight basis) are also shown. Obviously starch yields equivalent to those recorded for cassava could be achieved if 60 palms with 400 kg of starch could be harvested per hectare per year; however, it is likely that there is a strong negative correlation between yield per palm and palms per hectare.

Table 5. Estimated Yield of Sago Starch

Production system	Yield per palm kg	Palms harvested ha^{-1} a^{-1}	Starch yield t ha^{-1} a^{-1}
Wild stands	138	40-60	7-11
Semiwild groves	138	60	11
Cultivated groves	138	138	25

Source: Flach 1977

Flach's (1977) estimates of starch yield are higher than those I have quoted; they are based on estimates of 1 tonne of pith per palm with 18.5% starch which yields 185 kg starch, and on estimates of numbers of palms harvested per annum in three situations (Table 5). The value of 25 t ha^{-1} a^{-1} of starch

from cultivated groves is comparable with yields from cassava;
this is equivalent to a conversion efficiency of 0.4% which is
within the range 0.2-0.9% for other tropical crops (Table 3).
His estimate of total dry matter production of 53 t ha^{-1} a^{-1}
is about half the "realistic" potential yield and the highest
yields from perennial C_4 crops (Table 3); on the other hand,
it is similar to best yields from perennial C_3 crops. The
usefulness of these calculations depends upon the accuracy of
the estimates, but they do suggest that good yields may be
obtained. Moreover they are sufficiently encouraging to justify
more detailed measurements of yield and its components from
other areas in order to demonstrate the production potential
of starch from sagopalm.

Palm sago in some ways is like sugarcane: the economic
yield is stored in the stem which is a biological storage container. This has a number of consequences. Firstly, a price
must be paid in terms of energy (and hence carbon) to maintain
this storage organ; therefore like most woody or tree species
the yield potential is probably less than herbaceous plants
which do not have this high maintenance cost. Secondly, yield
will be directly related to the size (mainly height) of the
stem. Thirdly, the leaf canopy where the carbon is fixed is
continually elevated as the height of the stem increases.

Palm sago yield in the Wet Tropics may be increased by
improving the supply of mineral nutrients or water and by controlling pests and diseases; these solutions are out of the
scope of the present paper. Instead I intend to concentrate
on three areas: Leaf photosynthetic and respiratory characteristics; light interception; and harvest index.

Leaf Photosynthetic and Respiratory Characteristics

Dry weight increase results directly from the photosynthetic
activity of the plant canopy. The rate of canopy photosynthesis
(P_c) can be written (Charles-Edwards 1979):

$$P_c = \frac{\alpha I_0 \tau C (1-e^{-kL})}{\alpha I_0 + \tau C} - R_c$$

where α = light utilisation efficiency of leaves

τ = leaf conductance at the top of the plant canopy

k = extinction coefficient for light

L = leaf area index

R_c = canopy respiration

I_0 = solar radiation

C = carbon dioxide concentration of the air

I will discuss each of the parameters and indicate where I think there is, or is not, scope for improvement in sago yields. The solar radiation and carbon dioxide concentration of the air are environmental factors and not easily or economically controlled by man in field crops. Light utilisation efficiency (α) is different for C_3 and C_4 plants, depending upon temperature, being higher in C_4 and C_3 at temperatures greater than 30°C and the converse at lower temperatures (Ehleringer and Bjorkman 1977). However, there seems to be little variation among C_3 and C_4 groups and therefore little scope for improvement. The maximum conductance of leaves at the top of the canopy (τ) is directly related to maximum leaf photosynthetic rate; this is known to vary within both C_3 and C_4 plant groups but correlation between photosynthetic rate and dry matter yield is poor. In addition, attempts to increase growth rate and economic yield by selecting for high photosynthetic rates have been unsuccessful. Recent research has attempted to partition respiration into a component associated with maintenance of the standing biomass and another associated with growth (McCree 1974). There is good evidence that growth respiration is linked directly with photosynthesis and therefore any attempt to reduce it may well reduce overall carbon gain. As discussed before, tree crops like sago may have a higher respiratory load than herbaceous plants, which is the price of storing starch in living tissue of the stem. Selecting for lower respiration rates does not seem potentially useful, even though it potentially can increase net production and harvest index (v. Corley et al. 1971).

It is my view that breeding or selecting for superior leaf photosynthetic and respiratory characteristics in sago would be

premature and unlikely to lead to any gains in the short term,
if at all, especially if sagopalm is like oilpalm where there
is little genetic variation in net assimilation rate (Corley
1973a). On the other hand, there appears scope to improve light
interception, the remaining term $(1-e^{-kL})$ in the equation.

Light Interception

The term $(1-e^{-kL})$ contains two biological parameters: k,
the extinction coefficient, which depends on the canopy architecture (such features as angle, dispersion and orientation of
leaves), and L, leaf area index, which is a measure of the size
of the light absorbing green surface. Theory predicts that
canopies with low values of "k" resulting from erect, well-
dispersed leaves use light most efficiently and therefore should
give highest yields. This suggestion led to much discussion and
experimentation on the influence of leaf angle. Both theory and
practice have shown that, at best, increases in photosynthesis
and yield of 10-20% can be achieved for erect compared with
horizontal leaf arrangement. However, in most plants increases
are less than this because the effect of leaf angle is reduced
by other factors. Thus attempts to increase photosynthesis and
starch yield in sagopalm by improving leaf angle would not seem
warranted at this stage, particularly if sagopalm is like oilpalm
with a low extinction coefficient, 0.43, indicating erect leaves
and relatively good light distribution within the canopy (Corley
1973a).

On the other hand, attempts to improve and optimise the
size of the photosynthetic system (L) so that the maximum amount
of solar radiation could be intercepted over the whole year would
seem the easiest and potentially the most successful way of
increasing sago yields. Leaf area index depends upon leaf area
per tree, and the density and arrangement of trees. Excellent
work in this area has been done on the oilpalm by Corley and his
colleagues (Corley 1973b; Corley et al. 1971; Corley et al. 1973;
Hardon et al. 1973), and it would seem to me that the same
approach could be applied to sagopalm, e.g. using growth analysis
techniques they have shown that the optimum L for oil yield is

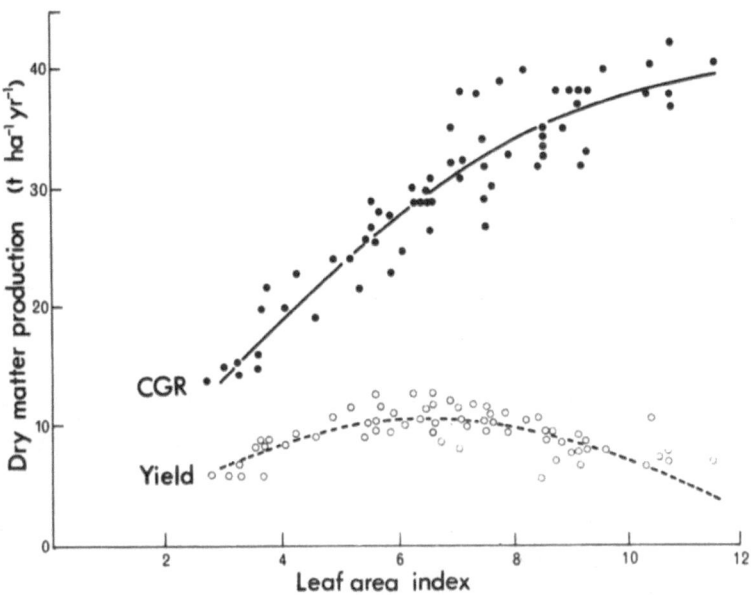

Figure 4.   Influence of Leaf Area Index on Crop Growth Rate
over a Year and on Fruit Yield of Oilpalm (Corley 1973)

lower than that for maximum vegetative production (Figure 4),
and they have identified harvest index as a critical characteristic influencing oil yield.

Harvest Index

It is well recognised that yield of economic product per
unit of net production or biomass, i.e. harvest index, is low
in "primitive" or "wild" agricultural plants, and that most of
the increase in yield of grain in many of our "developed" agricultural crops has been due to an increase in harvest index and
not in biomass production (Donald and Hamblin 1976), e.g. scientists at the University of Queensland have markedly increased
the grain yield of pigeon pea per hectare over that of Indian
cultivars by increasing harvest index. Thus in sago the aim
would be to increase starch yield relative to total plant dry
weight.

Flach's (1977) estimate of 0.47 for the harvest index of
sagopalm in Malaysia is comparable with many "developed" agricultural crops (Donald and Hamblin 1976). While this demonstrates

the potential of sagopalm, variation in harvest index obviously
exists when yields of starch per palm vary from nothing to 500 kg
(Barrau 1959). Thus there appears scope to improve yields by
increasing harvest index. The need to measure total biomass,
as well as starch yield in future is a prerequisite to any attempt
to assess how efficient the sagopalm is in converting solar energy
to chemical energy and attempts to improve sago starch yields.

Conclusions

As would be expected from a crop which has not been studied
in detail or subjected to breeding or selection, yields of and
efficiencies of conversion of solar energy of sagopalm generally
are low compared with more developed crops. There appears scope
for yield improvement by increasing annual light interception
through planting density and arrangement and by increasing harvest
index. The estimates of Flach (1977) suggest that in well managed
groves both high biomass and starch yields and adequate harvest
indices might be achieved. However, there may be religious,
cultural, social, economic or political considerations which may
prevent the full biological potential of sagopalm from being
achieved. This paper has been a theoretical treatment of the
subject of biomass production and sago yields, and by ignoring
these factors which may be overriding could well be naive.

REFERENCES

AUSTIN, R.B., G. Kingston, P.C. Longden and P.A. Donovan 1978.
    Gross energy yields and the support energy requirements
    for the production of sugar from beet and cane; a study
    of four production areas, J. agric. Sci. Camb. 91: 667.

BARRAU, J. 1959. The sago palms and other food plants of marsh
    dwellers in the South Pacific islands, Econ. Bot. 13: 151.

BROWN, R.H., and R.E. Blaser 1968. Leaf area index in pasture
    growth, Herb. Abst. 38: 1.

BULL, T.A., and K.T. Glasziou 1975. Sugar cane, Crop Physiology,
    ed. L.T. Evans; Camb. Univ. Press, Cambridge: 51.

CHARLES-EDWARDS, D.A. 1979. Photosynthesis and crop growth,
    Photosynthesis and Plant Development, ed. R. Marcelle
    (in press); Dr. W. Junk, The Hague.

COOPER, J.P. 1975. Control of photosynthetic production in terrestrial systems, Photosynthesis and Productivity in Different Environments, ed. J.P. Cooper; Camb. Univ. Press, Cambridge: 593.

CORLEY, R.H.V. 1973a. Effects of plant density on growth and yield of oil palm, Exp. Agric. 9: 169.

CORLEY, R.H.V. 1973b. Oil palm physiology; : a review, op. cit., ed. WASTIE and Earp: 37.

CORLEY, R.H.V., B.S. Gray and S.K. Ng 1971. Productivity of the oil palm (Elaeis guineensis Jacq.) in Malaysia, Exp. Agric. 7: 129.

CORLEY, R.H.V., C.K. Hew, T.K. Tam and K.K. Lo 1973. Optimal spacing for oil palms, op. cit., ed. WASTIE and Earp: 52.

DE WIT, C.T. 1965. Photosynthesis of leaf canopies, Versl. Landbouwho. Onderz. no. 663: 57.

DONALD, C.M., and J. Hamblin 1976. The biological yield and harvest index of cereals as agronomic and plant breeding criteria, Adv. Agron. 28: 361.

DUKE, J.A. 1977. Palms as energy sources: a solicitation, Principes 21: 60.

EHLERINGER, J., and O. Bjorkman 1977. Quantum yields for CO_2 uptake in C_3 and C_4 plants. Dependence on temperature, CO_2, and O_2 concentration, Plant Physiol. 59: 86.

FERRARIS, R. 1979. Productivity of Leucaena leucocephala in the wet tropics of north Queensland, Trop. Grassl. 13: 20.

FERRARIS, R., and G.S. Stewart 1979. Agronomic assessment of Pennisetum purpureum cultivars for agroindustrial application, Field Crops Res. 2: 45.

FLACH, M. 1977. Yield potential of the sagopalm and its realisation, Sago-76: Papers of the First International Sago Symposium, ed. K. Tan; Kemajuan Kanji, Kuala Lumpur: 157.

HARDON, J.J., Mokhtar Hashim and S.C. Ooi 1973. Oil palm breeding: a review, op. cit., ed. WASTIE and Earp: 23.

HATCH, M.D., C.B. Osmond and R.O. Slatyer 1971. Photosynthesis and Photorespiration; Wiley Interscience, New York.

LOOMIS, R.S., and W.A. Williams 1963. Maximum crop productivity: an estimate, Crop Sci. 3: 67.

LOOMIS, R.S. and P.A. Gerakis 1975. Productivity of agricultural ecosystems, op. cit., ed. COOPER: 145.

LOOMIS, R.S., W.A. Williams and A.E. Hall 1971. Agricultural productivity, Ann. Rev. Pl. Physiol. 22: 431.

LUDLOW, M.M. 1976. Ecophysiology of C_4 grasses, Water and Plant Life. Problems and Modern Approaches, ed. O.L. Lange, L. Kappen and E.-D. Schulze; Springer, Berlin: 364.

LUDLOW, M.M., and G.L. Wilson 1972. Photosynthesis of tropical pasture plants IV. Basis and consequences of differences between grasses and legumes, Austral. J. biol. Sci. 25: 1133.

McCREE, K.J. 1974. Equations for the rate of dark respiration of white clover and grain sorgum, as functions of dry weight, photosynthetic rate, and temperature, Crop Sci. 14: 509.

MONTEITH, J.L. 1972. Solar radiation and productivity in tropical ecosystems, J. appl. Ecol. 9: 747.

MONTEITH, J.L. 1978. Reassessment of maximum growth rates for C_3 and C_4 crops, Exp. Agric. 14: 1.

PURSEGLOVE, J.W. 1972. Tropical Crops. Monocotyledons 2; Longman, London: 426.

RUDDLE, K., D. Johnson, P.K. Townsend and J.D. Rees 1978. Palm Sago, a Tropical Starch from Marginal Lands; Austral. Nation. Univ. Press, Canberra.

STEWART, G.A. 1970. High potential productivity of the tropics for cereal crops, grass forage crops, and leaf, J. Austral. Inst. Agric. Sci. 36: 85.

WASTIE, R.L., and D.A. Earp (ed.) 1973. Advances in Oil Palm Cultivation; Incorp. Soc. Planters, Kuala Lumpur.

PRODUCTION OF LIGNOCELLULOSE AS A FEEDSTOCK FOR FOOD AND
FUEL PROCESSING

G. SIRÉN

CULTIVATED ENERGY SOURCE

The growing shortage of oil in combination with the rapidly
increasing price of processing oil products have evoked renais-
sance of interest in the production of domestic fuels in countries
short of fossil energy deposits. For instance, in Sweden attention has turned to existing substitutes for oil, i.e. peat and
waste from forestry, forest industry, agriculture and municipali-
ties.

In the long term however, these contributions seem insufficient to cover the demand for storable energy-rich organic
fuels. Thus a new concept, the cultivated energy resource, came
into the picture. Of the alternatives in question, wood fuel
produced by energy forests is one of the most promising; it is
in all respects independent of foreign imports.

The produced raw material, lignocellulose, suits however
other purposes quite well too. During World War Two cellulose
was used as fodder for ruminants. The storage of sugar was
easily overcome by processing cellulose to sugar. Today considerable quantities of protein-rich yeasts are produced in Finnish
factories. However, I am going to concentrate more on the fuel
aspect than on the food production for the reason that the energy
shortage seems to darken the future more than any other current
problems.

Of research projects involved in the production of biomass
for oil, fodder and food processing, the Energy Forestry Project
is the largest in Sweden. The project started officially 1976,
but preparatory work had been carried out since 1966; it belongs
administratively to the Swedish University of Agricultural

*W.R. Stanton and M. Flach (eds.), SAGO. The Equatorial Swamp as a Natural
Resource. Proceedings of the Second International Sago Symposium. All rights reserved.*
Copyright©1980 Martinus Nijhoff Publishers, The Hague/Boston/London.

Sciences, but is financed mainly by the National Board of Energy Research Development. This task was established by the Government in 1975 as a special unit of the Department of Industry. Of a total of 16 programmes the Biosystem-program is one of the largest.

The subprojects have their specified goals. The main aim of the whole project is to generate and collect knowledge about the potential of energy forestry and to produce the biological materials and methods necessary for large scale establishment of plantations in the event of a governmental decision in favour of oil-substituting domestic fuel production.

GENERAL PRODUCTION PRINCIPLES

In the Nordic countries, an average of about 0.15% of total incoming radiation, 700-800 kWh/m^2, during the growing season is photosynthetically converted into harvestable wood, 4-5 m^3/ha; this only corresponds to about 1.0 thermal kWh/m^2, and is far from the physiological potential.

If the photosynthetic efficiency is assumed to be in the range of 5.6-6.8% of the total available solar energy, physiological losses and a short growing season will reduce the potential production of harvestable wood; the foliage and root development requires about 40-50% of the assimilates formed. In Sweden these reductions indicate a theoretical annual maximum production of a magnitude of some 40-50 tonnes dry matter (DM) above ground per ha, provided that available solar energy is the limiting factor. For comparison, the theoretical maximum is of the magnitude of some 150-200 t per ha per year in the tropics, corresponding roughly to 60-80 t oil equivalents if used for heating.

In reality, however, there are a number of both abiotic as well as biotic factors which may impede the basic physiological functions, and thus reduce the production of the biomass. A few of the factors involved may be subject to influence, others are completely out of man's range of control. The climatic factors seem especially difficult to govern: light, precipitation, wind velocity, temperature, humidity, CO_2 content, and seasonal variation form a pattern of variables to which the biomass producing stand has to be adapted.

An informed selection of species and clones, a rational choice of spacing and harvesting age, optimising the foliage buildup and the chlorophyll content of the leaves, as well as the functioning of the stomata and the water nutrient uptake reveal some of the grower's possibilities to improve and increase the production. All these separate functions have their own optimum conditions, e.g. optimised water and nutrient uptake requires special soil structures, gas exchange conditions, pH range and a rather strictly composed nutrient solution always available in correspondence with the growth rate demand.

The main production principles thus include:

the buildup of climatically and pedologically well adapted biotopes of high-yielding plant species well fitted for biomass production purposes;

awareness and exploitation of the recycling phenomena of carbon, nitrogen and mineral nutrients, including skill to counteract growth reducing effects of controllable biotic and abiotic limiting factors; and

an optimum decomposition rate of organic matter for obtaining additional CO_2 supply to the atmosphere of the stand and continuous availability of nutrients in the rhizosphere during the growing season.

The main production principle for high production of biomass is apparent from the above. On any type of soil, peatland, abandoned arable land, even-surfaced forest land, etc., the first phase includes the establishment in a well functioning biotope. Thereafter the necessary ecophysiological steps follow for maintaining the nutrient circulation by compensating losses caused by harvesting or by unexpected leakage. By application of leaf area growth rate guided fertilisation principles, the risk of nutrient leakage remains low even in high producing stands.

SOME GENERAL PREMISES ON ENERGY FORESTRY

Among abiotic premises for energy forestry, well distributed precipitation is assumed to be one of the most important. In the case of Sweden, precipitation seems on average sufficient for maintaining the basic physiological functions at acceptable levels on the soil types available for high-yielding biomass production. Abundant precipitation is however not a categoric imperative, e.g. in arid or semiarid regions energy cropping can successfully be based on latex or oil producing xerophytes. The undisturbed functioning of stomata seems however to be one of the key premises for obtaining exceptionally high production values.

Standard broad-leaved energy forests should be located on areas with sufficient annual precipitation or otherwise available water resources in order to avoid high irrigation costs. Optimum uptake of water seems however to be a controversial matter. A premise for efficient root activity includes a rapid gas exchange in the rhizosphere, with sufficient access to water. In the Nordic countries, well balanced drainage, together with soil treatments adapted to the type and structure of soil, may produce the desired substrates for both stand establishment and for silvicultural optimisation treatments.

Another solution is to use plant species with efficient aerenchymatic root tissue for promoting the gas exchange between the rhizosphere and atmosphere. In this context, the O_2 : CO_2 relation of the soil should be investigated, especially in wetland stands, for finding out suitable crop species. That compaction by heavy harvesting machinery must be avoided is self-evident from the above; litter must be gradually decomposed for promoting recirculation of the mobile nutrients within the biotope established.

Methods and systems for appropriate nutrient supply need still be developed, although some experience from agriculture may be applied after modification. The land available in Sweden for energy forestry purposes consists mainly of abandoned agricultural land, sphagnum bogs, sedge mires and some forest lands.

One important biological premise for energy forestry seems to be the rigorous selection of extremely fast growing species and clones, characterised by a good capacity for vegetative propagation and coppicing, high net photosynthetic efficiency, efficient nutrient uptake, hardiness, resistance to disease, a high wood density, and clean and wellformed stem. For this purpose, special selection principles and testing programmes have to be applied.

Testing of Coppicing Clones
Clone selection

For obtaining maximum dry matter productivity, it seems necessary to operate not only with improved environmental conditions but also with superior clones selected for the biotopes in question. This means selection of fast growing clones potentially adaptable to a variety of natural and improved environmental production premises. (In this context, good growing conditions in Scandinavia mean an annual potential biomass production of the magnitude of 5 kg DM per m^2 of which 2.5 kg is harvestable).

The need for a standardised testing programme for fast growing coppicing species has arisen in order to facilitate international cooperation especially in energy forestry. In Sweden the type of available land has directed the interest in the first stage of short-rotation forestry towards the genus Salix. The suggested testing programme is structured as seen in Figure 1.

The initial material in this discussion consists of ramets that can propagate vegetatively. Since previous Salix material (primarily of imports from abroad) showed bad frost damage, the primary clone selection is now made as phenotypic selection on regional averages and from natural young stands often in suboptimal conditions. This decision was however preceded by an analysis of the problems involved in the principles of selection. The main questions were: should a selection be carried out as that:

of phenotype only or selection based on clone testing,
of phenotype based on control individuals or regional averages,
from natural or artificial stands,
from stands in optimal or suboptimal biotopes, or
from old or young stands.

Tests on quality

After primary clone selection, the preliminary test starts.

154

The principle is to multiply and homogenise the sample for further
testing and to document and compare the qualities of clones,
primarily the rooting ability. Certain information concerning
growth, resistance to parasites and pathogens, as well as climatic
disasters, can be obtained. However, initial results probably

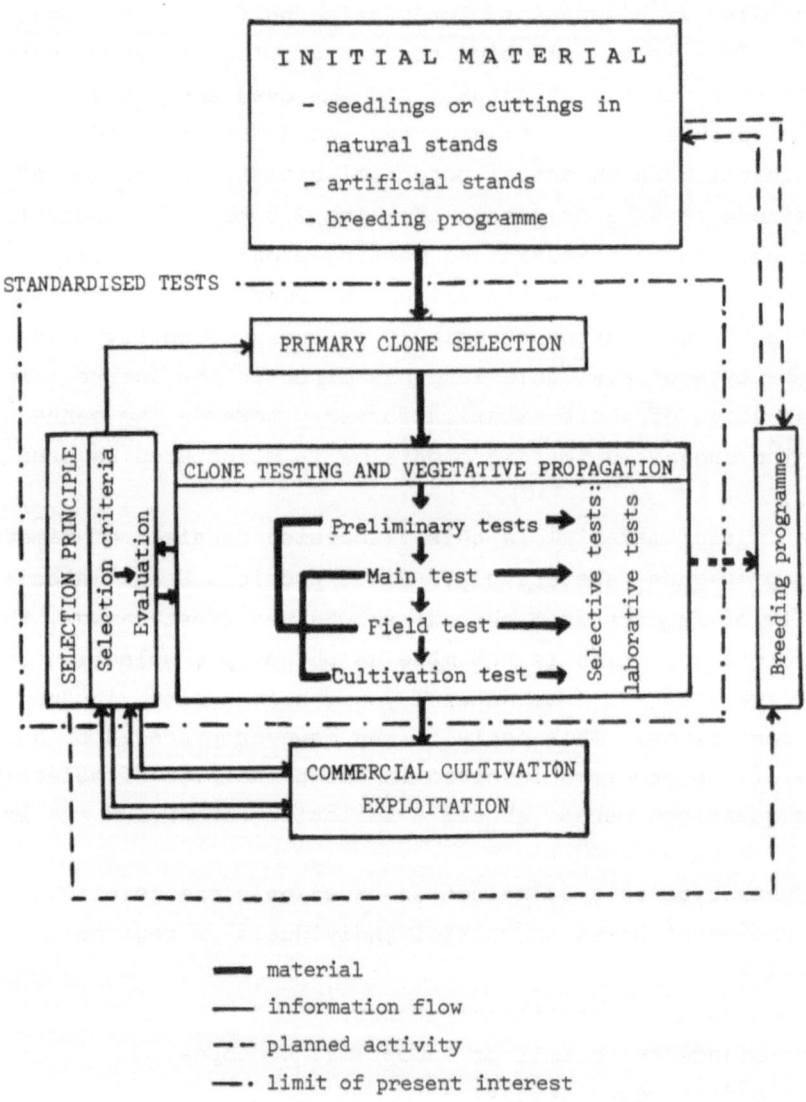

Figure 1. Guidelines for Standardised Test Programmes

reflect a large variation in the physiological attributes of the initial plant sample. This test lasts three years.

The clones which pass the preliminary test are used in the following main test. The different clones are now studied and compared with regard to several features.

Productivity and growth. The main tests were carried out on standard sites with limited environmental variation. Test design should be a randomised block with sufficient replications. The main test should be carried out in 3-5 years and will probably result in 5-50% of the clones being acceptable for productivity tests in the field; these are carried out on a variety of bio-topes. In order to investigate the occurrence of interactive effects between different clones and partly to determine future clone-mixing potential, a variety of different binary clone combinations should be included in the field tests. Production studies with nitrogen-fixation species should also be carried out.

In the field tests attention is directed to clone-specific responses to biological factors during establishment, management and harvesting of stands. Examples of factors important for practical cultivation are soil treatment, spacing, time and planting method, nutrition supply, watering, weed-control, harvesting methods, time of harvesting, rotation intervals, and the effect on coppicing capacity and survival.

Photosynthetic efficiency and nutrient minimisation are studied in laboratory tests. The result of these tests will enable a safe large scale use of the best clones in commercial cultivation for energy purposes, pulp industry, fodder or food production.

In Sweden, a countrywide survey was carried out in order to find a sufficient number of superior clones. The campaign was successful: 240 one-year old shoots were collected from mainly natural environments during the winter of 1978. In the spring the cuttings were planted; after shooting, the plants were watered as required, and nutrient was given on average at 2 g N/m^2 according to the leaf area growth during 8 weeks.

Pest and disease resistance: Weed control was carried out mechanically; no pesticides or fungicides were used in order to

get information about resistance. At the end of the growing
season the survival, growth rate, shooting ability, insect and
fungi diseases, frost damage, bud formation, volume and DM
production were investigated. The 24 best clones were selected
to form the preliminary basic material for further studies.

Sprouting capacity: After the rooting capacity of the
cuttings has indicated the establishment fitness of the clones
under survey, the shooting development is a parameter of main
interest. After a sparse shoot formation during the establishment
year, a more normal shooting pattern should be expected during
the second year - provided that the first year's shoots are cut
during the winter dormancy. A clear improvement in the sprouting
capacity was also observed in this case during 1979, well in
accordance with the earlier experience.

Differences obtained in total shoot length are predominantly
explained by differences in the competition situation and in
general differences. By excluding those clones which have bene-
fitted from less severe competition from the neighbours on one
or both sides in the row arrangement of the primary test, the
modified impact of the environment can be excluded.

Clone profile and comparative studies
A simple but comprehensive description of the test performance of the clones involved would be an advantage. In order to
create a standard norm, as well as norm for the superclones, a
concept of the so-called clone profile has been established. A
limited number of clearly discriminating parameters are chosen
for the purpose (in the example presented below, the bark thick-
ness is included as a non-discriminating parameter). A model
for the profile buildup of a few parameters is given in Figure 2.

The theoretical line formed by the extreme observations to
the right form the maximum theoretical superclone in the material
tested in this context; the average (not indicated) forms the
normal. Clones described by high values between normal and the
theoretical maximum for the most important characteristics, e.g.
rooting capacity, total shoot length, wood density, frost hardiness and total DM production, will most probably be selected for

further testing.

Comparison of consecutive clone profiles covering a range of years will certainly expose the critical moment when the profile becomes stable. Changing from comparisons within the clones to comparisons between the clones will, at least theoretically, disclose the point of commencement of stability, occasionally disturbed only by environmental disorders, or displaying ecological impacts worthwhile to include in the final profile of response of the clone in question.

In the pilot study carried out in Sweden (observations during two years hardly yield consistent results) only a few parameters offer possibilities for comparison over a 2-year period. This depends upon the fact that the autumn inventory has to be carried out in November-December. Because of the fact that height growth will cease in most of the clones in mid-September, the mean maximum height (calculated on stool basis)

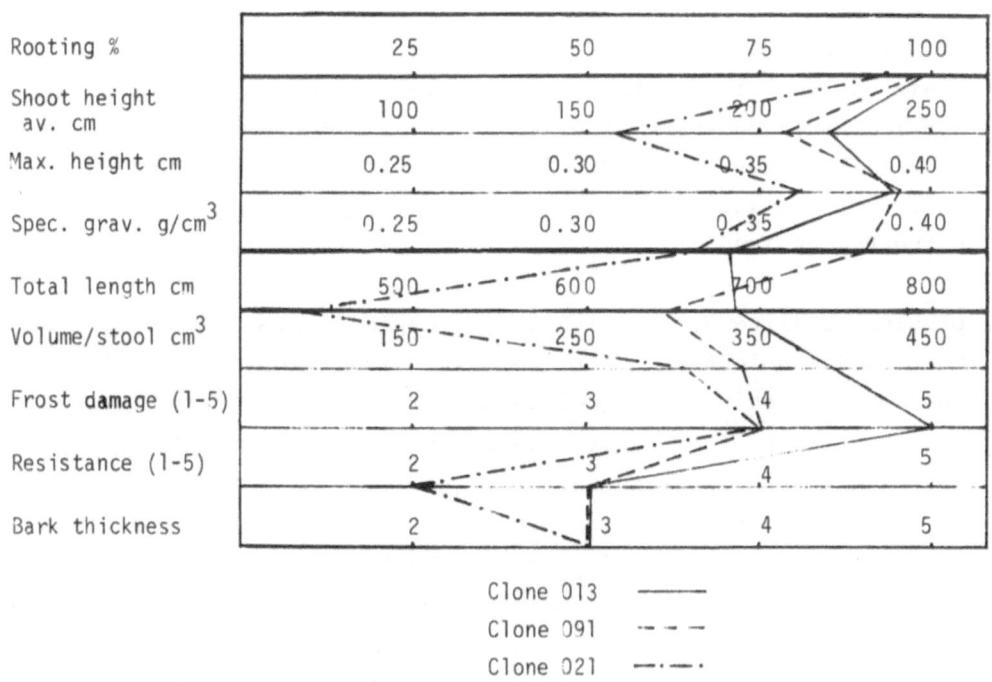

Figure 2. Examples of Three Clonal Profiles (Nov. 1978)

already affords a preliminary consistency test.

The present picture reveals that a few clones only have reached the stage of stability; many clones do not permit any serious decision making at this early juncture, and the proposed duration of the preliminary test, 3 years, seems reasonable.

It should be mentioned that the clones previously accepted for main tests and field tests did not reach the same mean height as the ten best of the clones included in the preliminary test described above. This is promising for the reason that the clones undergoing the main test produce, in the present field experiments, stem wood quantities indicating annual production rates of the magnitude of 25-30 t DM per ha; this corresponds in terms of thermal energy content to about 10-12 t of oil per ha and year. Thus the main goal of the Energy Forestry Project seems within reach as far as clone selection is concerned.

Experimental Results

Generally speaking, research work development in the field of biomass should concentrate on three types of activities: basic biological research, experimental, mainly agrosilvicultural application, and system analysis including machinery development.

In the case of the Swedish project the basic research was included at a late juncture (1978) in the project. The results hitherto obtained are therefore based mainly on earlier findings from related research. The field experiments have covered problems dealing mainly with selection and production conditions. The species selected so far are mainly from the genera Salix and Populus; this choice relates mainly to the land available and the results from pilot experiments. The genera Alnus and Betula are also represented.

Regarding Salix, selection has so far revealed the definite importance of climatic adaptation. Fast growing species from Central or Western Europe may produce acceptable 1-year crops but are unlikely to withstand the Scandinavian winter climate, allowing for 2 or 3 year harvesting intervals; only Scandinavian, Finnish, Siberian and Canadian clones seem hardy enough. Of about 2,500 selected clones, only 9-10 clones have passed the tests (1979).

The stand establishment experiments so far cover only about 9-10 different soil types. The treatment variables are conventional: drainage, soil structure, pH, fertilisation, cutting size, planting depth and spacing, establishment date, weed control etc. Results have been encouraging with a few clones.

The ecophysiological optimisation field experiments have so far only covered pilot studies regarding depth of tillage, water regime, nutrient supply, leakage, spacing, length of rotation, harvesting age, etc. The best results have so far been obtained from small experimental plots of size 16-25 sq m.

Net-area plots have produced 2.5-3.0 kg/m^2 (5 clones) with fertilisation programmes (Ingestad 1967, 1979) and of the magnitude of approximately 80-200 kg $N.ha^{-1}.yr^{-1}$ related to leaf area growth index. The total leaf area in one clone (Salix smithiana Q 666) attained a top value of > 20 m^2 per m^2 soil surface. In existing cutting orchards (0.3 ha), annual production has so far been in the range of 16-18 $t.ha^{-1}.yr^{-1}$. The most promising species are S. viminalis, S. aquatica, S. smithiana and S. purpurea so far.

Within the field of basic research some interesting results have been obtained concerning the relation between growth rates and the internal status of mineral nutrients. Different clones react in different ways when subjected to a stress of nitrogen; this can be seen both in growth rates and in rates of net photosynthesis. If these results are valid in a field situation, it means that the possibility exists of making early tests of the clonal material in the laboratory and using this information when selecting suitable sites and management of the individual clones. In future it is hoped to add an extensive physiological test programme to the more conventional one that is in use today.

In connection with the ecophysiological research programme, there is also research on wood anatomy and growth analyses. Since the recirculation of nutrients will be of extreme importance within "energy forests", as well as within any form of optimised biomass production, a large programme on litter decomposition and nitrogen turnover is included in the project. The possibility

of using biological nitrogen fixation to reduce the energy input
in terms of nitrogen fertilisers is also subject to intensive
studies.

ENVIRONMENTAL CONSEQUENCES

The main consideration concerning unwanted environmental
consequences appear at two different levels: in the system of
biomass production and in its use. During the production stage,
the main source of environmental impact is thought to be that
caused by leakage of nutrients, especially nitrogen. For reducing
the risk to an acceptable low level the nutrient supply has to
be adjusted to the plant's uptake capacity, e.g. using the leaf
area growth rate as a dosage guide. Continuous sampling of the
nutrient content of seepage water from a variety of sample plots
will reveal the facts about leakage in due course.

In Sweden not only nitrogen but also the heavy metals occurring as impurities in some types of fertilisers are the subject
of observation. Some of the species may prove to be excellent
collectors. Hydrological consequences caused by excessive water
uptake by high-producing crops seem theoretically possible. For
this reason, water regime studies are being carried out both
with lysimeter techniques and flow control methods. Changes in
the water pH is of considerable interest in Sweden where soil
acidification caused by airborne SO_2 is widespread.

The impact of energy forestry on the flora and fauna of the
land used as well as in surrounding land areas is also of great
interest. The high nutrient content of the energy stands will
most probably result in a variety of fungi and insects establishing themselves in the biotope. By rigorous selection, the clones
accepted for further production experiments also demonstrate
rather high resistance to both disease in general and to many
other pests during the main growing season. Biological control
of both insects and rodents is under consideration.

To reduce negative effects on the landscape in connection
with a possible introduction of large scale energy forestry,
comparative studies are being carried out by means of models.
To avoid restrictive impact on recreation and outdoor activities,

large cultivation areas are subjected to appropriate planning regarding pathways, ditches and channels. Generally speaking, the environmental consequences of energy forestry seem considerable but not impossible to master, at least in the production stage. On the other hand, the negative impact on the environment is predicted to be quite difficult to counter in the utilisation stage.

However, compared with for example oil, the SO_2 content of wood from energy forests is much lower. The problems arise from unavoidable noxious gases and polyaromatic carbohydrates (PAH:s) caused by inefficient oxygen supply to the burning gases. New types of reactors and ovens as well as new types of traps for inadequately burned combustion gases must be developed. However, this problematic aspect lies outside the present biomass production project. As a possible consequence, it deserves mentioning that the return of the ash as fertiliser probably will prevent certain soil depletion effects in the case of whole tree use; in that case the leaves should be used for food and fodder extraction.

In addition to the Energy Forestry projects, a few minor projects on reed, straw and marine production are being carried out in Sweden. For some time the use of municipal waste has received attention. Considerable interest has also been attached to the collection of slash and other residue from conventional forestry, which together with peat, straw and reed have a great advantage in that they are currently available. Potentially, these sources may substitute a total of 25% of the oil import; the remainder has to be covered by energy forests, direct solar energy and electricity.

HARVESTING

A special project is involved in the development of harvesting machines. A prototype for harvesting reed is operational today. For harvesting energy forestry crops, two prototypes representing different function principles are under development; stand geometry has to be adjusted to the functions of the harvesters. Direct chipping or bunching is one of the main problems.

A preliminary cost analysis reveals that chips are already competitive with oil in large areas of Sweden.

CONVERSION

The wide range of alternative end uses of wood-based biomass
has generated a large number of special projects on conversion.
The high amino content of the leaves has generated interest in
the field of protein production - the papermills are reluctant
for the time being, but a few initiatives indicate already a
change in the attitudes towards the short-fibre raw material.

From the energy point of view, the ethanol and methanol
projects seem both to be of outstanding interest, especially
the ethanol process based on fresh mixed raw material seems
promising; the methanol project has, on the other hand, already
reached a lab-semipractical scale. Swedish enterprises can
today offer well functioning conversion systems.

ENERGY BALANCES

Pilot analysis of the energy input-output of a production
system for heating purposes suggest that well planned intensive
energy forestry will result in a ratio better than 10 : 1 in
favour of output. The studies cover all work, materials and
machinery necessary in the operations, land reclamation, stand
establishment, production, harvest, chopping, local distribution
of chips and the reconstruction of furnaces and related equipment.
Further conversion may reduce the balance to some 5:1. However,
if N-fertilisation can be avoided by N-fixation symbionts in the
optimised biomass, the balance may easily be readjusted to the
former level.

ECONOMICS

The economy of the energy forestry will depend on the fluc-
tuations of the oil price. At present, the low annual inputs
seem economically favourable for the growers.

SOCIOECONOMIC CONSEQUENCES

According to the working hypothesis, large scale intensive

biomass or energy forestry will have certain effects on employment in rural areas. A plantation of 100-ha size is assumed to produce 700-1000 t of oil equivalent (toe) per year and to give full employment to 2-3 persons. Transferred to a nationwide scale, this figure implies profitable employment to 20-30 thousand persons on a total area of a million hectares. The net contribution to domestic energy supply would at best be of the magnitude of 70-80 TWh (thermal), with an oil substituting effect of about one quarter of the present import.

It is relevant to relate the general impact of biomass production on the CO_2 content of the atmosphere. There has been an extremely confusing worldwide debate concerning the CO_2 balance; it has been said that neither methods nor techniques exist today to counteract the release of CO_2 to the atmosphere caused by the burning of fossil fuels. The author is of the opinion that a considerable increase of the biomass production, by establishing large scale energy plantations as complements to existing forests, would be an ecologically safe, welcome and efficient countermeasure to the increasing CO_2 content of the atmosphere. Energy forestry may be the best remedy for the detrimental consequences of unwise use of the world's natural resources.

REFERENCES

INGESTAD, T. 1967. Methods for uniform optimum fertilization of forest tree plants, 14th IUFRO Congress Section, Muenchen.

INGESTAD, T. 1979. A definition of optimum nutrient requirement in birch seedlings. III. Influence of pH and temperature on nutrient solution, Physiologia Plantarum 46(1).

SIRÉN, G., T. Lestander and L. Sennerby 1979. Standardized procedure for testing of fast-growing species. A preliminary proposal, Energy Forestry Project Tech. rep. no. 2; Swed. Univ. of Agric. Sciences, Uppsala.

POTENTIAL AND PROPOSED DEVELOPMENT OF SAGO (METROXYLON SPP.)
AS A SOURCE OF POWER ALCOHOL IN PAPUA NEW GUINEA

E.B. HOLMES   and   K. NEWCOMBE

ENERGY POLICY

Papua New Guinea has developed a large and rapidly growing
dependency on imported petroleum products for commercial energy
supplies. Analysis of energy trends and evaluation of possible
scenarios for future energy demands indicated clearly that the
fuel requirements of PNG would continue to grow, whilst the global
capacity and motivation of oil producers to supply the market
would decrease. Therefore there is a great need to develop a
less energy intensive, more resource conserving energy policy
for the future of PNG (Newcombe 1979).

In this paper, PNG's proposed policies for the development
of renewable sources of energy are presented with particular
reference to sago (<u>Metroxylon</u> spp.). More comprehensive reviews
of policy are provided in Newcombe and Weick (1978) and the
Government's Energy Policy and related strategy (Ministry of
Minerals and Energy 1979).

RENEWABLE ENERGY SOURCES

Briefly, the sources of renewable energy in PNG which have
been considered on a national scale can be placed into three
main categories. These are:
Harvesting solar energy directly,
    e.g. photo voltaic cells, solar water heating;
Hydro-power,
    e.g. both macro and micro facilities;
Biomass resources,
    e.g. Forest residues and sawmill wastes, fuelwood cropping;
         broad-acre farming of cassava and sugarcane; nipapalm;
         sagopalm.

The estimated potential of these resources is summarised
in Table 1. Hydro-power, currently providing 2% of the nation's
energy, represents at least three times the anticipated energy
demands for all end-uses in the year 2000. Just 1% of the solar
radiation incident in PNG is about 260 times that amount and the
sum of biomass resources (Tables 1 and 2) amounts to 31-64% of
the estimated energy demand. Therefore, the renewable energy
potential of PNG is very large, being considerably in excess of
the anticipated energy demands of 2000 AD. But on what criteria
should the most appropriate patterns of energy source be chosen?

Table 1. Potential of Some Renewable Sources of Energy
in Papua New Guinea

Energy requirements for 2000 AD	1160×10^8 MJ
Hydro-power	4415×10^8 MJ
Solar energy	33×10^{15} MJ
Biomass resources	$359.5-622.3 \times 10^8$ MJ
Broad-acre cassava farming	11.6×10^8 MJ

Source: Newcombe 1979

Some guidelines for national development in PNG are provided
by the national development goals (CPO 1976); these include:

1) Balanced rural development,

2) Participation by small holders,

3) Equal distribution of goods and services, and

4) Self-reliance and independence.

Therefore, it is implied the energy sources chosen should
not have larger, concentrated facilities. Newcombe (1979) con-
cluded that the preferred pathways to energy self-sufficiency
were those that used naturally occurring renewable resources
requiring little in the way of conscious management to ensure
a sustained yield. Also, the technologies of energy conversion
should be as small in scale as possible and simple to construct
and manage. It is considered that the use of sago-starch for
ethanol fuel production is fully compatible with these objectives.

Table 2.   Selected Biomass Energy Resources

Biomass Source	Quantity Available	Energy Value[1] 10^8 MJ
Wastes at major sawmills	339,000 t	27.1
Residues and culls from major forestry operations	2,072,000 t	166.7
Combined sawmill and forest residues from small operations[2]	820,000 t	65.9
Sagopalm[3]	300,000 ha	60.4-323.2
Nipapalm[4]	47,000 ha	35.4
Existing fuelwood cropping[5]	1,500 ha (300 ha/yr)	4.0
Total		359.5-622.3

[1] Ethanol is regarded as the energy form produced from sago and nipapalms, and for the purpose of these calculations its energy value is taken as 85% of motor spirit, or 29.2 MJ/l, compared with its actual enthalpy value of 21.3 MJ/l.

[2] 34% of the resource available from major existing sawmills and forestry operations, estimated on a pro-rata basis using cubic metre input data to the sawmills (Office of Forests 1978).

[3] Area available for harvesting is estimated by Cavanaugh (1955) for the Sepik provinces, and estimated from maps provided by J. Zeick (1979, pers. comm., FPRC, Boroko, PNG).

[4] Only firm data available on the resource size is in respect of the Purari Basin. This is a very small portion of the actual resource which has not otherwise been surveyed.

[5] Firewood crops planted on the Wahgi swamp up to December 1978 (L. Martin, Provincial Forests Officer, Mt. Hagen, pers. comm. 1978).

SAGO RESOURCES

Sago is the starch accumulated in the trunk of the sagopalm. This starch is the main staple in diets of the indigenous population of equatorial swamp or lowland areas, being extracted by traditional means. In Papua New Guinea, sagopalms are found in large natural stands in the Sepik River and Gulf Province estuarine and freshwater swamp areas (Map).

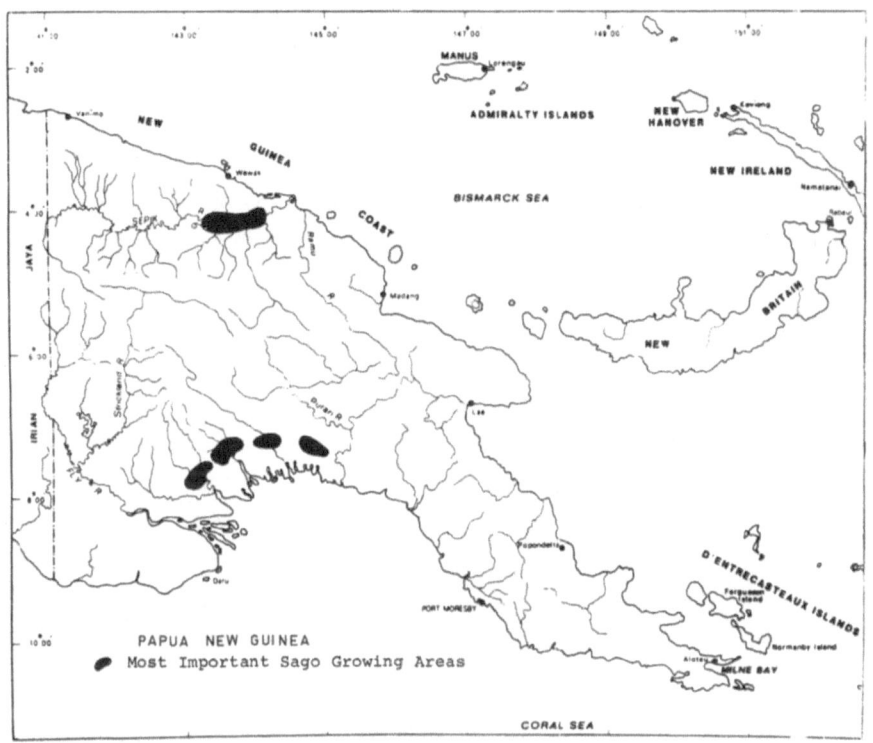

PAPUA NEW GUINEA
Most Important Sago Growing Areas

The palm genus, Metroxylon, consists of some six species, of which economically the most important is M. sagu; M. rumphii is found predominantly in the wild sago areas. The sagopalm is the dominant species in those areas of equatorial swamp in which it grows, thereby forming part of a specialised ecosystem; presumably, sago possesses a competitive advantage over other species in such growth conditions. The soils along the Sepik river are heavy clays high in organic matter, only flooded seasonally by the Sepik; this periodic flooding with water high in mineral elements should provide all the nutrients required for optimum sagopalm growth.

Sago can be propagated from seed, but once established is usually propagated by means of tillers or suckers arising from the trunk base. Therefore, there would be no need for continued replanting programmes, either in wild or cultivated sago areas.

Starch is accumulated in the trunk of the palm during the vegetative growth phase. In the Sepik, palms are usually harvested immediately prior to fruiting (12-15 year old palms)

which give the highest starch yields. No major studies of the
sago growth cycle have been completed.

Analysis of different classes may indicate either earlier
maturing varieties or the most suitable harvest times. But this
is not considered an immediate priority, as it is estimated that
there are sufficient resources in the Sepik and Gulf provinces
to supply starch to any proposed fuel alcohol industry.

RESOURCE POTENTIAL

Estimates of total sago resources in PNG are crude, for
although total plant distribution can be determined from aerial
photographs, yields of starch cannot. However, sago has been
investigated as a source of starch at least 11 times between
1953 and 1972. Such investigations have provided data on which
preliminary evaluation of the resource potential can be based;
the yield data are summarised in Table 3.

Table 3. Potential Productivity of Sagopalm in
Papua New Guinea

Yield logs/ha/yr	Dry starch kg/log	Dry starch kg/ha/yr	Ethanol l/ha/yr	Area for 2 million litres ha
Minimum 7.5[a]	154	1,155	693	2,899
Maximum 42[b]	164	6,888	4,133	484
Plantation 138[c]	185	25,530	15,318	78

Source: (a) Cavanaugh 1955
(b) Toyo Menka 1972
(c) Flach 1979

Cavanaugh (1955) estimated that there were one million acres
(about 454,500 ha) of sagopalm stands in the Sepik province, and
a further 1.5-2.0 million acres (0.68-1.6 million ha) exist in
the Gulf Province (Zeick 1979). Newcombe (1979) conservatively
estimated that throughout PNG there were at least 300,000 ha of
harvestable sago, basing his estimate on readily accessible stands.

Estimates of productivity in sago swamps in the Sepik region
vary considerably. Cavanaugh (1955) estimated that there were

7.5 mature boles per ha, yielding an average 154 kg dry starch (DS). Toyo Menka researchers (1972) recorded 40-42 mature sago boles per ha with an average wet weight of 1400 kg or 168 kg dry starch (assuming 20% wet starch at 40% moisture content). Thus yields of 1.16-7.06 t (DS) per ha could be harvested from wild stands of sago each year. It is important to note here that the Toyo Menka scientists have been the only group so far to systematically survey the population density of mature boles and their starch content in wild swamp regions of the Sepik Province.

Flach (1979) considered yields of 7-11 t (DS)/ha/yr possible from wild stands of palms in New Guinea and the Moluccas; under cultivated consitions, these yield levels would improve considerably. He estimated that in a well maintained plantation, 138 trunks/ha/yr could be harvested, each weighing approximately one tonne yielding 185 kg DS. Therefore, yields of 25.5 t DS/ha/ yr are achievable with present plant productivity.

Annual ethanol production can be calculated from the dry starch yields using a conversion factor of 0.6. On this basis, in wild sagopalm areas, 693-4234 l/ha/yr of ethanol could be produced. In between these estimates were data computed per ha per yr of ethanol to 930-1,380 l (Morris 1953), and 1,630-1,930 l (Edwards 1961, citing estimates from Office of Forests, PNG). However, all sources agree that following the first rotation and cleaning operation in wild sago areas, yields of starch would improve considerably.

Ethanol yields of 15,300 l/ha/yr could be computed from Flach's data (1979). Although this figure may not be achievable immediately, it represents a feasible target for ultimate yield. Taking the productivity estimates for wild sago, therefore the Sepik and Gulf Provinces sago resources could yield 207-1270 million l of ethanol per year; this represents 0.25-1.5 times the total petrol import for PNG (Newcombe 1979). Using the area provided by Flach (1979), managed plantations of sago over the same area would yield 4-6 billion litres of ethanol, or 70% of PNG's year 2000 energy demand. However, the actual potential will not be known until the industry is developed and incentive given for resource evaluation.

PROPOSALS FOR A SMALL SCALE SAGO-ETHANOL INDUSTRY

On the basis of the foregoing discussion, a 1-2 million litre pilot sago-ethanol industry has been proposed for the Sepik area. Sago-starch can be converted by enzymes, or acid hydrolysis into sugar for fermentation into alcohol. The technology for this process is well developed and it is considered that an energy selfsufficient means of converting cassava-starch to alcohol in small 1-50 million l/yr factories is feasible for sago processing. Fibre residues left after starch extraction and the woody sheath of the bole could either be dried and burnt, or the pith digested to produce methane to provide the total energy required of the process.

Site

The Sepik area was chosen because of the large accessible area of wild sago stands and the availability of land suitable for sagopalm cultivation, mainly floodplains. Sago gardens are planted and maintained either by individual or family groups and represent the main food source of the area. However, not all mature boles are harvested each year and it is believed likely that sufficient high quality sagopalms are available for ethanol production, without disrupting the local food supply. In preliminary surveys, cultivated sago logs had far greater total fresh weight and starch content than wild sago.

Estimates of the motor-spirit requirements in Angoram and Wewak are compatible to a 1 million litre per year industry, especially when the conversion of outboard motors and standing diesel electric generation to alcohol fuels is considered.

Prices of motor-spirit are high in the Sepik area, being 30 toea/l in Wewak and 35 toea/l in Angoram; further price rises of 3-5 toea/l are expected by mid-1980.* Therefore, for reasons of sago availability, motor spirit consumption and price, the lower Sepik river area is considered the most suitable region for the initial sago industry, even though the feasibility study to be

---

* 1 kina(k) = US$ 1.45
            = 100 toea (t)

conducted in 1980 will consider other locations in the West Sepik.
The favoured factory site will probably be at Toway, a village
downstream of Angoram and still in the midst of wild sago swamps;
the river will be used as a natural energy transport system to
bring sago boles to the factory or storage ponds.

Servicing of Industry

Provision of sufficient sago-starch is the major problem
for the proposed industry. Approximately 11,100-13,333 sagologs
will be required to service the industry each year. However,
due to difficulties in harvesting and transporting logs during
the dry season, up to 80% of the annual log requirement could
be delivered during the six month period of the wet season.
Therefore, pond storage facilities will be established above
the factory site. The effects of such storage on starch contents,
methods to improve storage, and alteration of harvesting patterns
to reduce storage periods will need to be studied.

An area of 484-2900 ha could supply the factory but only
131 ha are required for high yielding plantation sago. About
3000 ha of natural sago stands are available in a five-mile
radius around Angoram, and many times this resource is available
in the area around Marienburg-Toway. Although plant yields are
not critical to the servicing of the industry, it is obvious that
high yielding material would dramatically reduce transportation
and harvesting costs.

It is envisaged that logs will be harvested by local village
people and either floated to the river, or dragged there by oxen.
Sago logs will be accepted at the factory gate regardless of
origin and at a flat rate. A premium based on the starch content
and yield of the log would be added to encourage the sale of high
quality sagopalms.

In order to emphasise diversity and security in the procedure
for servicing the industry, additional feedstock acquisition
strategies will be adopted. Firstly, a long term lease will be
obtained on both sago swampland and Kunai grassland close to,
and upstream of, the factory. Planting of highly productive
sago cultivars would begin immediately at up to 200 ha/yr; these

logs would be harvested after 8-10 years.

In the interim, any deficit in sago supplies off the river will be made up by harvesting wild sago from the swamp areas under lease.

Cost of Alcohol Production

The direct costs of ethanol production have been calculated to be 12-15 toea/l, of which 4-7 toea/l represents the cost of logs and 8 toea/l (max.) for starch conversion. Allowing for blending, wholesaling and retailing costs of 7 toea/l and capital servicing and depreciation costs of 7 toea/l, a profit margin in the year one of 1-5 toea/l at 1979 prices is achieved. There-fore, given escalating fuel prices, the process is considered to be a fully commercial proposition as well as being ecologically sustainable.

PROPOSED SCHEDULE OF INDUSTRY DEVELOPMENT

It is hoped to design and have in full operation a 2 million litre per annum factory by early 1982. Therefore, a time scale for development operation has been developed (Table 4).

Table 4. Timetable of Development for East Sepik
Province Sago Ethanol Industry

Year	Operation
1979	Prefeasibility study
1980	Feasibility study Finance raised Construction companies selected
1981	Plant construction commissioned Sago plantations initiated, 100-200 ha/yr
1982 Jan.-Feb.	Alcohol production to 2 million l/yr
1983	Expansion to 5 million l/yr commissioned
1984	Production of 5 million l/yr
1986	Expansion to 10 million l/yr commissioned
1987	Production of 10 million l/yr

The pre-feasibility study proposed for 1979 includes an evaluation of:

1) Wild sago starch contents within the Sepik region
2) Total fermentable content of sago boles
3) Inspection of sago processing equipment
4) Designs for processing and distillation equipment.

During 1980, a full feasibility study will be undertaken prior to selection of management and construction companies. Amongst other items, a detailed evaluation of the social and biological impact of the project on the region will be made. It is not intended to disrupt substantially the present patterns of rural village life in the Sepik region or reduce the food resources of the indigenous population.

Following at least a one year assessment of the factories' operation, capacity and success, a decision will be made on expansion of the industry to 5 and then 10 million 1/yr by the end of 1987.

CONCLUSION

The Department of Minerals and Energy considers that sago-palm is a tremendous natural resource which can be developed to reduce or replace petroleum imports. It is a crop which is suited to lowland swampy areas, can be harvested throughout the year and can give high and sustained yields without expensive agricultural management. Therefore, sago can be used without harmful effects on the present ecological patterns and wilful disruption of the traditional patterns of life in Sepik.

Acknowledgements: The senior author would like to thank the Department of Minerals and Energy for providing the travel grant which made attendance at the Symposium possible.

REFERENCES

BARRAU, J. 1959. The sago palm and other food plants of marsh dwellers on the South Pacific islands, Econ. Bot. 13: 151-62.

CAVANAUGH, L.G. 1955. Sago Flour Production, Sepik River; Forest Products Res. Cent., Boroko: 10 pp. (mimeo.).

CENTRAL PLANNING OFFICE, Papua New Guinea 1976. The National
    Development Strategy; Waigani.

EDWARDS, E.T. 1961. The Natural Stands of Sago Palms, Metroxylon
    spp. in the Sepik River Area of New Guinea and their
    Possible Use as a Source of Commercial Starch; report
    submitted to the Board of Directors of Leo Fielders & Co.
    Ltd., Sydney: 26 pp.

FLACH, M. 1979. The sago palm: A potential competitor to root
    crops, 3rd Intern. Symp. Tropical Root Crops 1973, Ibadan:
    170-6.

MORRIS, H.S. 1953. Report of the Melanau Sago Producing
    Community in Sarawak; Col. Res. Stud. no. 4, London: 184 pp.

MINISTRY OF MINERALS AND ENERGY, Papua New Guinea 1979. Energy
    Policy and Planning for Papua New Guinea; Konedobu.

NEWCOMBE, K. 1979. Energy and Urbanization in Papua New Guinea:
    The industrial city of Lae; Papua New Guinea Human Ecology
    Programme Tech. pap. PNGE/T7, Cent. for Resour. and
    Environ. Stud., Austral. Nation. Univ., Canberra: 53 pp.

NEWCOMBE, K., and L. Weick 1978. Energy in Papua New Guinea's
    Future; Policy and Planning Unit, Dep. Minerals and Energy,
    PNG.

TOYO MENKA KAISHA Ltd. 1972. Feasibility Study for the
    Establishment of a Sago Flour Plant in the Territory of
    Papua New Guinea; Forest Products Res. Cent., Boroko: 34 pp.

GLUCOSE OR GASOHOL

W.R. STANTON

FOCUS ON SAGO

The sago industry of Southeast Asia is at present depressed, moribund, or an almost extinct part of the agrarian systems of most of lowland Southeast Asia. The author's interest in the history and resuscitation of the industry springs from two lines of investigation.

Firstly, in the 1960s when FAO was interested in closing the so-called protein gap, a gap which has subsequently been proved to be largely a misunderstood "food gap", he investigated various methods, suitable for operating at the village or household level, of converting carbohydrate biomass into protein by fermentation (Brook et al. 1969). This was to be an extension to, or substitute for, the methods of increasing the available cheap protein derived from plant protein crops (Stanton et al. 1965). The investigation led to his devising simple methods for the conversion of starch into protein by a solid substrate fermentation (SSF) derived from the TEMPE process (Stanton 1972). As with many apparently simple fermentation processes, these methods have become more complicated and sophisticated in their development towards greater protein conversion efficiency and a higher degree of food safety in the end product; Dr. Raimbault presents these subsequent developments. These new ideas have added significantly to the commercial feasibility of the technique of SSF processes, though there is still scope for development of the original concept of a static fermentation.

Secondly, arising from the search for sources of starch, the author was asked to report on the sago industry in Sarawak and to suggest how it might be rehabilitated and its products

Grimbledon Down

Courtesy of Mr. Bill Tidy and the New Scientist 78/1099 (1978) (upper), 86/1202 (1980) (lower).

made competitive with those of the cereal and root starches.
This work culminated in the First International Sago Symposium
in 1976 when viable solutions to energy crises were dimly visible
(Stanton 1977a), and the enthusiasm for the production of microbial protein from hydrocarbon sources had not yet dissipated
(Abbott 1974). Thus, the early years of the author's study were
in relation to the food and industrial uses of starch.

However, in the past few years, techniques using renewable
resources to alleviate the energy shortage have become a principal
feature of the economic programmes of many countries and, from
the point of view of agriculture contributing to energy supply
via biomass production, the concept of energy budgeting as a
feature of the design of agricultural systems has been developed
to a fine art (Greenfield and Nicklin 1979). The author realised
that the renewal of sago production might be given a new twist,
in that this palm presented a number of useful features that could
change the current opinion on ethanol (fuel alcohol) production,
namely, that of regarding cassava, sugarcane and sugarbeet as the
ethanol crops par excellence.

At no stage in the development of the author's own thesis
was it envisaged that energy, and particularly ethanol from
biomass, might prove to be the panacea for a world starved of
energy. But, as many eminent students of photosynthesis have
shown (Hall 1978, 1979), the sun pours abundant energy onto land
suitable for agriculture. Even with populations of the size
envisaged for the 21st century, they have shown that this energy
is in excess of that required at the present level of demand
(Brown 1980), though whether it is generally feasible or not to
gather the biomass arising from fixing solar energy is a different
question. There is also the question of food versus fuel (Clarke
1980).

Nevertheless, as has been shown by Bruin (1980) for example,
although solar energy is trapped by photosynthesis in both agriculture and forestry, much of the energy trapped in the latter
is unavailable, as forests have a protective function for the
environment, and for this reason the energy latent in the biomass
may not be harvested. Further, there are various restrictions

on applying good agricultural land to energy cropping. The criterion for deciding on the proportion of the land which may be applied to this purpose is discussed later. This is a choice for each individual country to make; for some, energy farming is inappropriate.

In spite of the gross difference in climate between them, there are yet similarities for environmental management between the cold-temperate/subarctic swamplands and those of the equatorial environment. The approach to the harvesting of the solar energy falling on these northern swamplands, as proposed by Sirén (1978) and his colleagues, is quite different from that proposed in the sago system, but nevertheless it is an alternative energy biomass producing system which is appropriate to the theme of this Symposium, the equatorial swamp as a resource.

THE HISTORY OF THE USE OF BIOMASS

Man has traditionally used plant biomass for food, structural material, clothing, craft and fuel. The advancement of agriculture can be stated to be, in terms of nett increase of production per unit area, in proportion to the adoption of new agronomic methods, the development of tools and machines, or the chemical approach e.g. fertilisers, insecticides, fungicides and herbicides. However, the current argument is that the main increase in production per unit area has been in terms of the increased energy applied to agricultural production (Lovins 1977; Green 1979). So great has been this increase that, for many crops today, the energy applied to the crop is greater than that derived from it, notwithstanding the fact that the crop itself has been for many months assimilating solar energy. The staple crops selected by man, and the processes applied to them after the harvest for the preparation of food, have been those crops and processes which yielded the greatest energy from the food, unencumbered by bulk, weight, water or extraneous non-digestible matter.

Carbohydrates such as sugar and starch, and also the cellulosic crops, cotton, sisal, abaca (Stanton 1977c) with the exception of woods, are examples of biomass energy trading, having little adverse nutrient-depleting effect on the soils of the production area, since the extracted carbohydrate does not take along with it essential mineral nutrients and therefore does not, incidentally, decrease the productivity of the producing area. This contrasts with the situation which pertains for many other traded plant products, which not only export the product of photosynthetic energy but also a component of mineral nutrients. This hidden loss is being increasingly recognised in the plantation industry in tropical agriculture and thus, for example, in the palm oil industry, efforts are now being made to return all the waste materials derived from palm oil extraction back to the land. Palm oil, as exported, is a biomass energy source without the concomitant export of nutrients.

CHOICE OF ENERGY-CAPTURE PLANTS

Current staple crops, such as the potato, the true and aroid yams, cassava and the tree starches became domesticated as food plants because of the ease with which a relatively low-fibre food energy source could be obtained from them. These starchy roots and stems are easy to transport, fairly easy to store, or could be left in the ground until required, they are also easily processed. Starch is one of industry's most useful feedstocks, and the question arises as to why the swampy equatorial lowlands of the old world have not become extensively cropped with sago, and indeed why the culture has not extended to monsoonal swamps, in the light of the crop's utility and versatility.

The argument has been put forward that sago and similar palms take too long to establish for these perennial food "trees" to be considered for cashcropping. This is a shortsighted view. The agronomic potential of the carbohydrate palms is largely untapped, the architecture of a starch palm as a solar energy capture and storage device (silo) is elegant, especially in comparison with tubers and cereals as starch storage organs. Trunk storage is bulk trading, tuber storage is retailing, the starch in cereals is almost an incidental commodity. The argument against palms which invokes superiority of one photosynthetic pathway over another, the C_3 vs C_4 controversy, is debatable in the context of the ecology under consideration.

In the author's view, this failure to develop tree starches has also been due to the adverse effect on commodity trading in starch of the ability of America in the past century to put cereal starches on the world market very cheaply. To the benefit of the developing countries, we are seeing the end of this era. An additional effect in relation to sago culture has been the local rise in social preference for rice, in spite of the recognition of its unreliability as a food source when compared to sago (Hamilton 1959), and in the cocoa/oilpalm era, the competition from these crops for the sago lands (K. Tan, pers. comm.). That these events have not been without repercussions has been shown by the recent comments by the Malaysian Minister of Agriculture on the state of West Malaysian lowland cultivation, in which the existence is noted of 1-2 million hectares of "idle" land, the alleged causes being drought, labour shortage and the low returns on rice.

COMPARISONS BETWEEN FORMS OF FIXED PLANT-CARBON

In considering here the competition between carbohydrates, the goal of "a useful industrial-carbohydrate feedstock" is in mind. Unlike the aim of the 1979 UNIDO Conference in Vienna, or that of the 1977 International Symposium on Alcohol Fuel Technology in Wolfsburg, the objective of this Symposium is not confined to the production of ethanol. From the point of view of elevating equatorial swamp areas to a renewable resource-based industry, it is recognised that the harvested carbohydrate may at the factory be more beneficially converted to some other product than ethanol, hence the adoption at this Symposium of the phrase "food or fuel".

An advantageous feature of ethanol as a feedstock for industry is the intermediate step of distillation. This step, regarded solely as an industrial chemical separation technique, permits the separation of the preferred component of the fermentation from other non-volatile pollutants, thus allowing the use of a crude precursor. However, advances in separation technique are becoming increasingly sophisticated so that this unique feature of distillation, a separation process, does not have the appeal which it had formerly.

In reviewing the range of carbohydrates available and considering their relative freedom from extraneous components, it is appropriate to compare the properties of cellulose, starch and sugars, both from the points of view of plant production and as convenient feedstocks. The costs of production and preparation are inseparable for assessing the total process economy, since what may be gained by low in situ cost of production in the field may be counterbalanced by a series of relatively high costs for transport, cleaning, disintegration, separation and digestion, crude ligno-cellulose being the prime example of a low in-field cost linked to a high preparation cost (Ghose 1979). On the other hand the field product may contain nutrients of value to the fermentation, thus reducing the cost of preparation, cane juice and molasses being examples.

Ligno-cellulose is nature's contribution to the supply of reinforced plastics. It is resistant to disintegration and separation into its component polymers, lignin and crystalline cellulose; even the component molecules themselves are resistant to further breakdown, viz. both the crystalline cellulose and the lignin cement are extremely resistant to breakdown by biochemical or biological means. These properties are well recognised by the wood-pulp and paper industries. It is for this same reason that those research workers who have opted for ligno-cellulose as the starting point for their carbohydrate based fermentations have had a preference for accepting, as raw material, the finished product of the paper-pulp manufacturer. Using processed pulp as the starting point overcomes or, more correctly, obscures the principal problem associated with the preparation of cellulose feedstock, i.e. the energy involved in preparation; paper-pulp manufacture is a high energy-consuming process.

The biochemical advances achieved in synthesising mutant microorganisms in the course of some 40 years, for the enzymic degradation of en masse ligno-cellulose have been pathetically poor. This contrasts with the advances made in enhancing conversion efficiency, through microbial manipulation, in other processes requiring starch-degrading enzymes, citric acid producing

enzymes and the even more dramatic increases achieved for improving efficiency of complicated processes, such as the multiple
enzyme systems required to produce antibiotics or modified sterols.

This difficulty in overcoming the resistance to degradation
of nature's most fundamental building block is understandable in
the light of a long period of natural selection, in the course
of which ligno-cellulose has become the preferred structural
carbohydrate. Nature built trees to last, even under tropical
rainforest conditions. Unlike starches, sugars and fats, nowhere
is ligno-cellulose found as a biological energy store; it is
always a structural or protective material.

Starches

It was for the above reason that the author took one step
down in the hierarchy of plants to locate those tree-like forms
which employed other carbohydrates than cellulose as their structural materials. At first sight it may seem strange to consider
the starch grain as a suitable structural material, but the starch-
containing palms demonstrate that, in conjunction with a fibrous
matrix occupying relatively little bulk and only a small proportion of the dry matter, starch grains packed in thin walled sacks,
parenchyma cells (the sandbag analogy), act as building blocks
enabling the palm trunk to withstand compression and bending
strains. Nature has combined this bio-pylon with a water and
fireproof biota resistant coating, the thick bark. These two
properties, a log-like package of starch and a rotproof packing
case, are now exploited in the all-the-year-round harvesting,
rafting (the cheapest way of transport) and in-water storage at
the factory of the harvest. This harvesting strategy meets the
economic desiderata of full use of expensive machinery by insurance of a regular supply of feedstock to the factory. An incidental feature is that the residual sugars in the starch, through
the reaction of lactobacilli, act as a preservative medium for
the starch column, a process which is well known to the people
of the Pacific region for their "clamping" preservation method
for sweet potato and taro, Ipomea batatas and Colocasia esculenta,
as 'poi' (Allen and Allen 1933), and in the now almost forgotten

practice of ensilage of the potato, Solanum andigenum, in Europe.

A problem associated with the production of ethanol from starch crops has been that although they provide the carbohydrate at high density and therefore reduce handling problem to the minimum, they do not furnish adequate ancillary fuel, in contrast to the sugarcane crop, for operating the factory, although bagasse is too bulky to be an ideal fuel. For the tree starches this lack of factory fuel does not apply because the fibrous bark, which is brought to the factory along with the starch, provides just about enough fuel for the disintegration, fermentation and distillation processes, the main energy demand being the last named unit process.

A hidden increment of fuel, available from all starch but not so readily from sugar crops, is that which could be provided from the wash water by converting the otherwise polluting liquor to fuel via a methanogenic fermentation. This has given rise to the question of whether it may in fact be a better strategy for a sago growing and processing complex to reduce the carriage of extraneous material to the minimum to the factory; viz. starch + other plant parts + water would be separated in situ in the field, transporting only starch milk by pipeline. This might be the preferred method where water transport of the logs presented a seasonal problem. The residues would be returned directly to the palms as mineral nutrients; the bark would be used in situ as fuel for the rasping and starch extraction machinery. Drying the bark for fuel would not, in the author's opinion, be a problem. Engineers like their machinery grouped together, but the precedent for splitting is already set, e.g. the modern combine-harvester is a complex mobile factory. Similarly the starch factory might be barge mounted; the tin industry in Malaya is based on barge mounted factories. The idea would not therefore be alien to local business.

Sugar Crops

In contrast to the starches, whilst sugar crops can give a high return per unit area of carbohydrate ready for fermentation without further treatment, they suffer from acute problems of seasonal variation in production and difficulty of storage of the crop; much extraneous bulk, water and cellulosic residues, is transported to the factory. Of necessity, most of these residues have been put to secondary use either as fuel for the factory, production of additional electricity for sale, or as a ligno-cellulose feedstock for other industries. But, as mentioned previously, it may be wasteful of agronomic effort to produce ligno-cellulose in the light of the postharvest energy required for processing. Sugarcane competes with abaca (Musa textilis) in a total biomass processing strategy; the latter's very high quality fibre, in comparison with bagasse, contributes a second readily marketable product. The extractable biomass concept is so new that the plant kingdom has yet to be surveyed for all the species to which it might be applied.

To summarise, the virtues of sago and the other starch palms, in comparison with other biomass sources, have been obscured by the well established starch and sugar industries; attempts to alter the situation have been frustrated by the plantation establishment on the one hand and the inability to "look beyond tomorrow" on the other. This long range vision refers to the lag in commencement of production, a handicap which is argued elsewhere, and which is unfairly applied to these little known tree crops with their plant breeding future - cloning, polyploidy exploitation, genetic engineering, in contrast to sugarcane with its plant-breeding past.

The Market Product

To the point of presentation of the starch logs to the factory log pound, very little of the original fixed energy is lost, the minor starch-preserving lactic fermentations hardly making any demand on the store of energy derived from photosynthesis. The crown of leaves, left in the field, is a small proportion of the fixed carbon bulk, but is a valuable sink for the in-plant-mobile essential minerals, particularly magnesium and potassium, which would need to be returned to the field if the crown ligno-cellulose were transported to the factory. Thus, almost the whole of the photosynthate could be marketed as starch itself. This direct sale is an option open to every sago production area and gives rise to the question implied in the title of this paper.

Starch, glucose, resynthesised biopolymers, modified sugars, organic acids are all allied products demanded by the market. The alcohol fermentation is not unique, even if "fermentation" is regarded as an alternative primary step (Ribeiro Filho 1979; Yakovleff and Goharel 1979; Yamazoe 1979). Most of these alternative products of fermentation are at present imported into developing countries. This market option is mentioned again later in consideration of the actual fermentation.

CHOICE OF TECHNIQUE: THE GASOHOL JUNGLE

In natural product technology there are instances where there are no feasible process alternatives from which to choose to obtain a particular product. The chemical technology of ethanol production is not one of these. It suffers from the burden of thousands of years of awareness of the complex sequence of techniques of enzyme production, saccharification, mashing, wort separation, fermentation and distillation. Until almost the middle of the present century alcohol production was recognisably a classical art dominated by the brewer and wine maker; it has still this identity in the minds of the policy makers and the engineers associated with project approval and realisation respectively.

Thus, as revealed at the Vienna (UNIDO Sec. 1979) meeting on gasohol, the different techniques which can now be amalgamated to provide a system for alcohol production, the different degrees of sophistication, the differences in technologists' opinion of "what is the objective in terms of a fuel?", must bewilder the administrator, and hence delay progress in selection and implementation of projects. In the present context this simply means delaying the decision to plant sago, a decision which needs to be made now, whilst the administrator is still surveying the energy options which will become available more than a decade into the future. Part of the current nightmare of the fermentation engineer is the fear that a core feature of a complex he has designed may be out-of-date by the time the complex is operational, the giant tower fermentor of ICI being the current example in the news (Senior 1980).

Changes in features of peripheral unit processes may not affect the overall design so much. They may justifiably be changed as a routine feature of replacement in the plant maintenance programme. Other processes may be modified whilst the plant is running, or refinements (automation, computer control and feedback) interposed, but the dilemma remains for core elements such as fermentor design. Care in selection of plant for the core process is therefore particularly important.

Feedstock Preparation

Fortunately, with sago it is not necessary, for the sake of the economics, to invoke the question of whether one should hydrolise the cellulosic residues, as is currently the question with bagasse and sweet sorghum (Kelly 1979). However, the starch has to be converted to glucose at present and for this operation the choice is between the following methods of feedstock preparation:

(a)   Chemical hydrolysis
(b)   Enzymic hydrolysis
(c)   Microbial hydrolysis
(d)   Physical disintegration (a recent development from Canada 1980)
(e)   A combination of above techniques

Until quite recently, except for endproducts in which the organoleptic (and for potable spirits mythological) quality of the endproduct was important, chemical hydrolysis of the starch was preferred for cheapness and is still used for many fermentations, including for instance the glutamic acid fermentation. However, the cost of enzymes, and the efficiency with which they can be made to operate on the substrate with the aid of immobilisation, have now made enzymic hydrolysis competitive (v. Baker on current developments).

Why, however, should one hydrolyse at all when one might use a dual organism fermentation, e.g. the Symba process? Alternatively, by genetics and genetic engineering one might introduce amylolytic activity into an ethanol producing yeast. Can this latter transformation be done without impairment of the other properties of the cell, that is, without adverse effects on the core property of the ethanol-synthesis pathway, the intra-cellular associated pathways, or such other features as flocculatability (Prince and McCann 1979), high metabolic temperature optimum and the capacity for self-optimisation, whereby infection can be tolerated and media sterilisation is unnecessary (Meyrath 1979)? These are all questions still under research at the present time.

The Fermentation

Economic optimisation of plant and process dictates reduction in the volume of the plant by shortening of the reaction time, increasing the efficiency in the transformation of feedstock to product and optimising the yield of product in relation to pre-cursor. This raises two considerations. Firstly, can the rate at which the metabolic processes within the cell be accelerated by improving the rate of supply and withdrawal of metabolites? In practical terms this improvement in reaction rate can be effected in various ways (Jackman 1977):

1) By improving the perfection of mixing (the mixing quality, e.g. the CHEMAP spherical fermentor);

2) By improving the rate of removal of the product (Vacuum stripping of Finn and Boyajian 1976);

3) By increasing the tolerance of the organism to high concentrations of feed (osmophilic yeasts);

4) By increasing the tolerance of the organism to higher concentration of product, as an alternative to product stripping (this goes beyond feedback repression to the condition where the cell is resistant to ethanol toxicity) (Kodama 1970).

Secondly, are fermentors obsolete? Carrying enzyme engineering to its logical conclusion, one can envisage the whole biochemical process from feed to product being operated without any reproducing microbial cells actually in the reaction mixture. The "hundred per cent inoculum", Melle-Boinot technique, concept is a member of the hierarchy of immobilised enzyme systems. It is the one lying closest to the system with living reproducing cells, and it is debatable whether it is to be regarded as an enzyme engineering or a microbial fermentation technique (Stanton 1975).

Differences of opinion exist on the relative merits of maintaining high concentration of a flocculant yeast by natural settling, as compared to returning a nonflocculant yeast to the fermentor after "cleaning" by centrifugation; both techniques have merit (UNIDO 1979). The natural settlement method avoids the energy cost of centrifugation, but on the other hand centrifugation allows precise control of the liquor content of the slurry and the characteristics of the particles returned to the fermentor. The control is advantageous in relation to continuous fermentation and liquor stripping.

The above argument is part of the argument for sophistication as to how far one should separate the cell growth, respiration, enzyme producing activity from the biochemical reaction which produces ethanol. These considerations affect fundamentally the design of the core element of the plant. Fortunately the dilemma can be to some extent avoided, the decision set aside, by applying the depreciation rule; but the administrator who opts for a "fully proven in operation" design may later face embarassing questions from his board (his decision on the particular design having been influenced by pressure from salesmen) as to why he made the particular choice. This dilemma is largely due to the rapid rate at which energy prices and energy production techniques have changed in recent years. The ultimate choice in transportable fuels is of carbon vs hydrogen, nuclear and solar "energy" being convertible to "fuel" in the absence of carbon by the hydrogen pathway; this hydrogen alternative needs to be taken into account in making decisions on synthesising "carbon" based fuels (Schlegel 1978). There are many technical difficulties, none insuperable, to realising a hydrogen economy and it has the advantage of not being dependent on land/water capable of growing plants.

Product recovery

Product recovery is always an expensive component of fermentation processing. Much research has been conducted on reducing the cost. With alcohol the product might, theoretically, be stripped from the fermented mash in several ways (* all in Hartline 1979):

(a) Differential volatility (distillation: the classical method)

(b) Sieving (membrane and ultra filtration: Gregor,* Riley*)

(c) Adsorption/absorbtion and subsequent release of the alcohol (a form of gas chromatography: Miller,* using natural and synthetic zeolites: Baughman/Gelo*)

(d) Chemical combination, prior to separation, and subsequent recovery of the alcohol from the combination (at present only a theoretical concept: Lonsdale*; but the dehydration technique using polyphosphoric acid also belongs in this class of processes)

(e) Affinity volatilisation (eutectic mixture separation with liquid CO_2: de Filippi*); dibutyl phthalate: Pennsylvania U.*

(f) Solvent extraction (affinity of the alcohol for a second liquid immiscible with water, from which liquid it may be subsequently stripped, examples: Hexane stripping, Gasoline stripping - without subsequent separation: Chambers*)

(g) Vacuum stripping

(h) A combination of the above techniques.

As a commercial method at present, recovery is confined to distillation prior to fractionation of the fermented mixture, leaving a relatively concentrated highly polluting "slops" which is disposed off as animal feed, burnt for the ash, or dumped on land or in waterways. The last option is unacceptable for the very large operations envisaged for the future, and the second is wasteful of the fixed carbon components of the mixture. The above method also creates difficulties in avoiding fouling of the distillation equipment, particularly where molasses has been used as the feedstock. Partial clarification, by settlement or centrifugation, is effected in the processes which recycle or dispose of the yeast, though the liquid phase retains suspended solids and large dissolved molecules, or colloidal components which remain in the liquor after this treatment. "Bright" clarification, the objective of beer and wine manufacture, is not an economical option.

Provided that the fermentor can cope with a fibrous wort, suggested by Rudolph (UNIDO Sec. 1979), there is merit in leaving the fibre in during fermentation. Many residual solids are removed in beer and wine manufacture (a further example of the influence of craft on alcohol fermentation technology) because of their organoleptic side-effects, not because of their effect on the yeast fermentation. The plant fibre is not digested in the fermentation, but "cleaned" of trapped and adhering carbohydrates. It is beneficially retained for its function in helping clarification of the fermented mash prior to distillation; it would interfere with heat transfer, if left in the stillage. Its presence may make stirring of the fermenting mash more difficult, but this difficulty can be overcome in the design of the fermentor (v. Dr. Raimbault).

To summarise, just as modern farming (renewable resource production) has been accused of being glorified neolithic farming, so has alcohol fermentation and recovery yet to shake off the habits of middle eastern and oriental village brewers and distillers. Currently, recovery is confined to distillation, but several alternative methods are in the offing.

Plant Design

In contrast to the potable alcohol industry, the industrial alcohol fermentation industry is served by a handful of suppliers, though the number has grown in the last two years, as interest in the product has increased. There are certain principles, the consideration of which may help in making a decision on design.

Capital vs maintenance cost

Even though in the individual fermentors a batch process may be operated, the equipment generally will be required to run continuously, at least on a weekly basis. The production scheme envisaged by Kelly's (1979) approach requires for its economic viability long-term continuous operation and, in the author's experience, this is a very difficult goal to achieve even here in Malaya, where there is a well established plantation

industry with a developed engineering infrastructure. It takes
more skill to keep poor equipment running than good equipment.
The lowest tender, or the cheapest version of a design may not
be the most economical. Experience in the replacement of stain-
less steel by plastics has been varied, especially where long-
term continuous operation is required.

Most "offers" for installation of turn-key plant are to the
suppliers' own design to process the amount of feedstock, or
produce a given quantity of endproduct as specified by the client;
thus quality is linked to a patented design. The latter may
have been specified by the overall contractor for the plantation
and factory scheme. The most efficient fermentation system may
not necessarily be the most reliable, but two trends are visible.

Firstly, the era of cheap, low-efficiency, short-life fermentors is drawing to a close as labour costs and construction
material costs rise. Worldwide experience of running high efficiency fermentors is increasing and this experience may be transferred rapidly to developing countries. The latter do not want
second-rate technology, or equipment with high maintenance costs.

Secondly, although continuous fermentation only occupies
a small percentage of the installed fermentation capacity at
present, it is making steady progress as the preferred technique.
Maintenance of the desired inoculum, overcoming the problem of
infection, is now a feature of technique on which there is substantial experience and the main resistance is not technological or
economic, but sociological, due to the fear of directors of risk-
ing employing new designs and a new breed of technologist, whose
art they do not understand.

Labour requirement
    A strength of the cane-sugar industry lies in the degree
of automation which has been achieved in the field in land preparation and planting, tillage and harvesting. By contrast the
whole development of a sago industry is still in the early experimental stage, though in the epithet "logging for starch" is
implied a low requirement for labour per unit of starch handled.
Labour requirement for cassava and the state of development of

its agronomy are intermediate between the above two examples.
At present, in the author's view, the present architecture of
the cassava plant and its tubers is unsatisfactory for full
mechanisation, and the plant breeding policy in relation to
cyanide and phenolics also needs revision. However, once the
different starch feedstocks reach the factory the problems of
managing and operating the fermentation and distillation plants
are communal. Here the questions for the designer are:

Will it be cheaper, assessed over the life of the plant,
to have automation, or manual control?
Can automation be installed later?
Should one design for future automation, but employ manual
control at present?

The answers are not simple. During recent years research
workers in the design of fermentation plant have shown that,
even in the hands of dedicated and skilled operators, one must
accept that in a factory a high degree of perfection cannot be
attained all the time in all the technical and supervisory labour
force; significant (20% or more) improvements in performance can
be achieved by automated control of parameters and installation
of deviation-sensitive feedback circuitry, whether hydraulic,
pneumatic or electrical. One should not think only in terms
of the last named; it can be the least effective under the
physical conditions under which the fermentation plant is required
to operate. However, advances in robustness, reliability and
ease of maintenance of on-line measuring instruments, especially
those involving electrodes, have been dramatic in the last few
years. Precise control of pH, redox-state and specific ions
can have significant effects on the yield of product, and excessive economy in instrumentation is unwise.

Cheap labour and the so called intermediate technology associated with it can, as both China and India have found, prove to
be a developmental deadend. Since with sago one is planning for
an industry with a plantation cycle-time of 40 years or more, it
should be planned in the context of a likely change in the status
of labour. This change is already recognised in the forward
planning of harvesting and processing techniques of the oilpalm

fruit and rubber latex in Malaya; a radical change in use of
labour has already taken place in many areas where the cane
sugar industry operates. It may be assumed that a similar trend
in the attitude to labour will occur in the large equatorial
coastal swamps areas of Latin America and West Africa, which
are potential areas for sago production. Nigeria has developed
a very successful small still for village use (Stanton 1977b),
but this was designed for producing good quality potable alcohol
and could not be adapted to the scale of output required to
produce economical fuel alcohol.

Marketing Alcohol

The final stages in the preparation of alcohol are rectifi-
cation and dehydration. For the production of a fuel, the former
operation is an unnecessary added cost and, where the purchaser
is using the material as a basis for chemical industry, it is
likely that he will wish to undertake the final preparation himself. Only when the industrial alcohol industry becomes large
and well established will it become a proposition to market
special grades of alcohol.

Much has been written on the drying of ethanol for gasohol
mixtures in which the main constituent is petroleum (80-90%).
In consequence new methods of drying have been developed, including the use of starch itself as the drying agent. Paradoxically
the motoring press has often advocated adding water to fuels.

In the author's opinion, the goal is industrial-grade 95%
ethanol powered engines. Already the motor industry is working
on the commercialisation of these. Engines relying solely on
industrial alcohol have been shown to be economical and trouble-
free in long term test; for these drying the alcohol is not
necessary. An added selling advantage is that the product is
a lead-free "petrol", and already countries such as Sweden are
advocating legislation for countrywide lead-free motoring (New
Scientist 1980). For urban areas subject to smog, including
newly developing smog areas such as Kuala Lumpur, the clean
alcohol fuel, with its absence of residual hydrocarbons in the
exhaust, is obviously advantageous.

Industrial Chemicals Based on Ethanol as Feedstock

	Pathway	Primary product	Secondary product	In use
ETHANOL	direct use	$+ O_2$ H_2O ——— Ethylene glycol		
		steam cracking — Acetone		
		Butanol		In (3.8)
		$- H_2O$ ——— Ethyl ether		B, J
		$+ C_2H_2$ CO ——— Ethyl acrylate	⎫	J, It
		(or CH_2 CH CN, H_2SO_4)	⎬ Plastics	
		$+ C_2H_2$ ——— Ethyl vinyl ether	⎭	US, WG
		$+$ HCl Ethyl chloride		
	via ethylene	polymerise ——— Polyethylene ——— Polythenes		In (13.0), B
		$+$ Cl ——— Vinyl chloride——— P.V.C.		(B) projected
		$+ O_2$ ——— Ethylene oxide——— Ethylene glycol		B
		+ H_2O		
		$+ C_6H_6$ ——— Styrene-polymerise— Polystyrene		In (21.0), B
		$+$ HCl ——— Ethyl chloride ——— Tetraethyl-lead		
			Food manufacture	Widespread
			$+ C_2H_2$-Vinyl acetate P.V.A.	B
		$+ O_2$ ——— Acetic acid——— Acetic anhydride	Widespread
			$+$ Cl-chloracetic acid	In (3.7), B
			$+$ Ethanol-Ethyl acetate	B, In (4.9)
	via acetaldehyde US (120.0), UK, F Unspecified end products	condense ——— Acetaldol——— Butadiene / Butyleneglycol	It, C / B
		(via croton aldehyde) ——— Butanol	
			2-ethyl hexanol (by condensation + H_2)	B
		$+$ HCHO——— Pentaerythritol		
		——— Cellulose acetate		
		$+$ Ethanol——— Butadiene ——— Polymers		
		$- H_2O$		

Key:

B = Brazil, C = China, F = France, In = India, It = Italy, J = Japan
US = United States of America, UK = United Kingdom, WG = West Germany
Number in bracket: thousand tonnes

Some denaturing will be needed (Forsander 1979); however, this is a cheap process in comparison with the above mentioned separation processes. The need for this is an added reason for large-scale controlled production, but security is needed in many industries. The risk of misuse of the distillate should not be cited as a special problem inhibitory to the development of the alcohol industry; the drug addicts will consume the most bizarre compounds, and society cannot be deprived of all the otherwise useful products on their account.

There is also a potentially large market for industrial ethanol for rural lighting and heating, replacing kerosene and earth-gas (natural gas), in addition to the industrial chemicals, including the plastics market. In the Table are cited what are currently considered (UNIDO Sec. 1979) to be the more important products of the heavy chemical industry, for which ethanol is a feasible or the preferred feedstock. It will be noted that these include ubiquitously used plastics such as polythene, polystyrene, acrylic and vinyl resins, plasticisers, solvents, surface active agents, anti-freeze compounds, hydraulic equipment liquids, agricultural chemicals. Examples of the quantity (thousand tonnes) of some of these chemicals produced annually from fermentation ethanol in India are given; other countries such as Brazil, Japan, France, West Germany, U.K. and Italy also employ fermentation ethanol as feedstock for processes mentioned here.

As Yakovleff (1979) points out, the alcohol chemical industry is less scale dependent than the petrochemical industry for some endproducts, an example being the viability of small (10,000 t/yr) plants for producing ethylene from ethanol. For developing countries further advantages of the ethanol based industry are:

Close ties with agroforestry-products used as raw material, thus promoting agroindustry;

Ease of processing and therefore simplicity of equipment and in training personnel;

Flexibility in location of factories arising from flexibility in scale, more so than ethanol for fuel alcohol; and

Relative ease of maintenance of equipment, an important

consideration in many developing countries.

Those who argue that land capable of yielding a food-like biomass such as starch should only be used for food are raising highly debatable economic questions and, if the market is there, the technique of using starch as a fuel or industrial feedstock will be applied to starch processing. The ethanol process is not solely a fuel producing process however. As far as possible all nutrients are recycled, and a food can be retrieved from the system in the form of yeast, to be further processed to protein concentrates.

Ethanol:

is non-poisonous, relatively non-inflammable, non-explosive product;

is non-polluting if spilled, burned or used in cars;

does not suffer acute evaporation losses during transport in comparison with methanol, for example;

is easy to transport and store and has an indefinite storage life; and

is a relatively concentrated form of fuel energy and gives a long engine life.

Bulk transport and storage could not give rise to a Torrey Canyon or a Third Mile Island, or even the normal organic-matter pollution arising from the animal or vegetable industry.

CONCLUSION

The author suggests that sago-type societies were the dominant societies of insular equatorial, as distinct from monsoonal, Southeast Asia in the past, with unique sedentary guarded-tree economies that had evolved in an agriculturally unpromising part of the world. The sago eating societies may be grouped amongst the homostatic societies of the world in terms of the stability of their plant resource base, and therefore should be studied from the point of view of how man must live with his environment for the future (Lewis 1980); homostasis is inevitable unless one still believes in a future of superabundant energy resources from space and therefore a superabundant mineral and food supply.

The need for adopting homostatic recurrent resource using

systems is a criterion for the future, unless we allow "the greenhouse effect" to overtake us and let the world become so polluted that very few organisms will live; amongst the survivers we should find the palms and the blue-green "algae" (Bailey 1979; Laurmann 1979). The observed environmental tolerance of palms augurs well for our postulated application of their productivity to providing the renewable resources for industry, since history has shown that successful economic plants possess weed and colonisation characteristics, and a resistance to environmental disturbance. Tolerance of soil acidity, root anaerobiosis, flooding and salinity are the particular attributes of the sago-palm. The proposed new industry based on sago is unique as an industrial chemical industry, in being a positive contribution to conservative ecology. Sago is a Southeast Asian survival kit and the region's contribution to the reservoir of crops for the 21st century (Nagato and Shimoda 1979).

REFERENCES

ABBOTT, J.C. 1974. Single Cell Protein; Acad. Press: 25-45.

ALLEN, O.N., and E.K. Allen 1933. The Manufacture of Poi from Taro in Hawaii with Special Emphasis upon its Fermentation, Hawaii Agric. Exp. Stat. Bull. 70.

BAILEY, K. 1979. Sensation, not science; Comment. New Scientist 30 Aug.: 642.

BROOK, E.J., W.R. Stanton and A.J. Wallbridge 1969. Fermentation methods for protein enrichment of cassava, Biotech. and Bioengineer. 11: 1271-84.

BROWN, W.L. 1980. Food or fuel: New competition for the world's croplands, World Watch (in New Scientist 35(1199): 904).

BRUIN, S. 1980. Biomass as a source of energy, op. cit., TNO.

CANADA WEEKLY 8(17), 23 Apr. 1980: Cheaper alcohol produced.

CLARKE, R. 1980. Food or fuel, New Scientist 86(1203): 161.

FINN, R.K., and R.A. Boyajian 1976. Preliminary economic evaluation of the low temperature distillation of alcohol during fermentation; Process design and product recovery, paper presented at 5th Intern. Fermentation Symposium, Berlin, 1976 (abstr.).

FORSANDER, O.A. 1979. Misuse of alcohol from automobile fuels and preventive measures, op. cit., UNIDO (ID/WG 293/25).

GHOSE, T.K. 1979. Symposium on Bioconversion of Cellulosic Substances into Energy Chemicals and Microbial Protein,

Process Biochem. (March): 1.

GREEN, M.B. 1979. The energy balance of pesticide use; Energy and pesticides, Biologist 26(3).

GREENFIELD, P.F., and D.J. Nicklin 1979. Synthetic fuels - the interaction between energy accounting and cost accounting, Proc. Austral. Inst. Engineers Nation. Conf., 1979: N31-3 (mimeo.).

HALL, D.O. 1978. Solar energy conversion through biology - Could it be a practical energy source? Fuel 57: 322-33.

HALL, D.O. 1979. Solar energy use through biology - Past, present and future, Solar Energy 22: 307-28.

HAMILTON, A.W.H. (ed.) 1959. Malay Pantuns; (5th ed.) Eastern Univ. Press, Singapore:
"Jangan harap kepada padi, Mari tanam dalam gembia,
Jangan harap kepada mimpi, Hari siang manalah dia"
(Gembia, or sago, will always be available)

HARTLINE, F.F. 1979. Energy efficient alternatives to distillation could make fermentation plant materials an attractive source of liquid fuel, Science 206: 41-2.

JACKMAN, E.A. 1977. Distillery effluent treatment in the Brazilian National Alcohol Programme, Chemical Engineer. (Apr.).

KELLY, F.H.C. 1979. Cost control factors in the production of ethanol from sugar cane, op. cit., UNIDO (ID/WG 293/15).

KODAMA, K. 1970. Saké yeast. The Yeasts, vol. 3.: 225-74.

LAURMANN, J.A. 1979. Market penetration characteristics for energy production and atmospheric carbon dioxide growth, Science 205: 896.

LEWIS, C.W. 1980. Alcohol and methane - two biological fuels for man; Biofuels, Biologist 27(2).

LOVINS, A.B. 1977. Soft Energy Paths: Towards a Durable Peace; Friends of the Earth and Ballinger, Cambridge, Mass.

MEYRATH, J. 1979. New developments in continuous alcoholic fermentation intensification-simplification-economization, op. cit., UNIDO (ID/WG 293/8).

NAGATO, I., and H. Shimoda 1979. The present state of sago production and its future, Japan. J. trop. Agric. 23(3): 160-8.

NEW SCIENTIST 85(1193), 7 Feb. 1980: 384 - A lead-free Sweden hangs in the air.

PRINCE, R.G.H., and D.J. McCann 1979. Fuel alcohol from crops by continuous fermentation, paper presented at the Inst. Chem. Engineer., Sydney, 1979: 7.

RIBEIRO FILHO, F.A. 1979. The ethanol-based chemical industry in Brazil, op. cit., UNIDO (ID/WG 293/4, mimeo.).

SCHLEGEL, H.G. 1978. Hydrogen biology and hydrogen economy, State of the Art: GIAM and its Relevance to Developing Countries, ed. W.R. Stanton and E. DaSilva; Kuala Lumpur: 324.

SENIOR, P.J. 1980. Single cell proteins and advances in biotechnology, op. cit., TNO.

SIRÉN, G. 1978. Energy plantations, Sweden Now 6: 17.

STANTON, W.R. 1972. Improvements Relating to the Fermentation of Cassava: UK Patent no. 1277002/1972.

STANTON, W.R. 1975. Scope and application of immobilized enzyme systems in India and other developing countries, IFS/IFIAS Workshop on Socio-Economic and Ethical Implications of Enzyme Engineering; Poona, 1975.

STANTON, W.R. 1977a. The purpose of holding a symposium on sago, Sago-76: Papers of the First Intern. Sago Symp. "The Equatorial Swamp as a Natural Resource", ed. K. Tan; Kuala Lumpur.

STANTON, W.R. 1977b. Microbiological conservation of wastes with special reference to their application to developing countries, paper presented at the UNEP/FAO Seminar on Agro-Industrial Wastes 1977 (4/10) Rome: 29 (mimeo.).

STANTON, W.R. 1977c. Survey of Agricultural and Agro-Industrial Residues in Selected Countries in Africa, papers presented at the UNEP/FAO Seminar on Agro-Industrial Wastes; Rome (Ref. UNEP/FAO/ISS. 4/07): 29 pp. (mimeo.).

STANTON, W.R. 1979. Protein enriched food, the long road to commercialization, Proc. Symp. on Protein Rich Food in Asean, 1979; Malaysian Inst. of Food Tech., Kuala Lumpur: 109-13.

TNO 13th Internat. Conf.: Biotechnology - a Hidden Past, a Shining Future; Rotterdam, 1980.

UNIDO 1979. Workshop on Fermentation Alcohol for Use as Fuel and Chemical Feedstock in Developing Countries, Vienna, 1979.

UNIDO SECRETARIAT 1979. Review of Papers (ID/WG 293/43, mimeo.).

YAKOVLEFF, P., and M. Goharel 1979. Agro-chemistry comes of age; Fermentation alcohol as basic raw material for a chemical industry, op. cit., UNIDO 1979 (ID/WG 293/3, mimeo.).

YAMAZOE, A. 1979. Use of ethyl alcohol as chemical feedstock, op. cit., UNIDO (ID/WG 293/5, mimeo.).

COST CONTROL FACTORS IN AGRO-ETHANOL PRODUCTION

F.H.C. KELLY

ETHANOL

Ethanol has been used to fuel internal combustion engines
for almost a century, during which time there has been a succession of shortages or fears of shortages of crude oil-based
supplies. Two distinct peaks occurred around the turn of the
century and during the 1920s; a third peak occurring during
World War Two had somewhat more localised reference.

Various agricultural crops have been used as a source of
ethanol depending upon local circumstances. During the latter
part of the last century the upsurge in interest was related
more specifically to overproduction of crops of potatoes or
sugarbeets in Germany and France. At present Brazil has embarked
on a comprehensive production programme based on sugarcane which
also has its economic relationship to world sufficiency in sugar.

In the U.S.A. the word "gasohol" has been coined to describe
alcohol-petrol blends currently being the subject of user-
acceptance studies especially in the state of Nebraska. The
author used a fuel of similar type quite regularly in his own
car during the years 1938-1945 under the trade names of 'Shellcol'
or 'Plumecol' as marketed in Queensland.

The alcohol for blending was produced from sugarcane molasses
which at that time in Australia was otherwise virtually unsaleable except in very restricted quantities. The oil companies
were very reluctant to market the blend and did so only under
act of parliament, which act still remains on the statute books
but has not been enforced since about 1946.

The blend employed at that time was for 5% ethanol, but
under the exigencies of wartime restrictions an endeavour was
made to increase this to 15%, a move which was unsuccessful

*W.R. Stanton and M. Flach (eds.), SAGO. The Equatorial Swamp as a Natural
Resource. Proceedings of the Second International Sago Symposium. All rights reserved.
Copyright ©1980 Martinus Nijhoff Publishers, The Hague/Boston/London.*

under tropical wet season conditions and resulted in two phase separations. Present day dispersants may be more effective and up to 20% blends have been reported to have been used successfully in several localities.

Around 1940 the author was marginally associated with moves to extend ethanol production using sugarcane itself as the raw material, with a price of M¢ 11 per litre being the target figure under consideration. Both the retail price of sugar and the retail price of petrol have increased by an average annual rate of 3.2% since that time (if peaks are evened out) or an overall increase by a factor of 3.5 in 40 years.

COST REQUIREMENTS

The cost of energy forms a substantial proportion of all basic costs in any community, thereby emphasising the importance of keeping these as low as possible. The author has closely watched the projection and achievement of costing in the field of nuclear energy since the end of World War Two, a costing exercise always "to be blest" as the source of abundant low cost energy, as soon as other forms of energy reached a certain cost barrier. Some thirty years later and several cost stages higher the barrier remains still tantalisingly just ahead. For agro-ethanol to make a useful impact on the national economy it must be low cost.

Although ethanol is useful as a chemical feedstock as well as for fuelling jet engines or either spark or compression ignition internal combustion engines, it is for use in the petrol fuelled engine that greatest interest has been generated. It is therefore necessary also to obtain an answer to the question: What is the real cost of petrol?

The retail price of petrol is invariably masked by charges and surcharges of various types employed by Government or even semi-government bodies for a source of revenue, and the price which the ordinary motorist pays bears little relationship to the real cost of petrol.

The author has, by numerous routes, attempted to obtain an

answer to the question just posed but with limited satisfaction. One rule-of-thumb which he has found to be moderately successful is that the retail price of petrol without taxes is about 2.5 ($\pm$ 0.5) times the price of the crude oil from which it was produced; thus from a crude priced at US$20/bbl a retail price of M¢ 67/1 would be indicated. This is currently considered to be a high price to pay for energy, and figures around one third of this would be much more acceptable for national economic planning as well as for the purse of the private motorist.

The question then to be asked now becomes: Is it feasible to produce ethanol at a price of the order of magnitude of M¢ 20-25 per litre? Low cost for producing any commodity implies high efficiency, low labour input, i.e. capital intensive, high productivity and large scale operation. Most of these require-ments are incompatible with many of the experiences in developing countries.

Ethanol is one commodity which can be produced on a small scale with elementary technology from a wide variety of naturally occurring phytological materials, and has in fact been produced in this way since prehistoric time. Experience has shown however that socially this is a highly dangerous undertaking with many undesirable consequences. Because of the very great difficulty in controlling production of this character, it is almost a necessity to eliminate small scale technology as a viable line of development.

Component Costs

In order to be able to identify high cost components in the production cycle of agro-ethanol the author has found it helpful to divide the general overall flowsheet into eleven subunits going as far as the storage of ethanol at the distillery site; to this needs to be added transportation and other distri-bution costs. This procedure also has the advantage that varia-tions of even ± 50% with respect to individual items only margin-ally affect the overall cost, and the possible scope of such variations in each sector can be individually examined.

Numerous comparisons have been made concerning potential

productivity of plant crops as agro-ethanol sources. When making these comparisons each source has both favourable and unfavourable conditions and it is necessary to be quite specific when dealing with such situations, e.g. it is not uncommon to compare best productivity conditions for a new crop with average conditions for the same crop grown elsewhere without examining the requirements for best conditions, or the likelihood of the new crop achieving less than best conditions under real-life large scale situations.

For successful development of an agro-based ethanol industry it is essential to apply the best of 20th century technology at each of the agricultural, harvesting, transporting and processing stages. The industry would not provide a panacea for unemployment, the minimum skill being that of a driver of heavy mechanical equipment or the general supervision of computer-controlled processing.

To be in business to supply liquid energy at containable costs it is necessary to adopt the attitudes and procedures of successful international energy corporations. It is necessary to think like an oil company, to employ technology of a similar degree of sophistication and to maintain a research and development organisation of corresponding magnitude and skill, which means recycling up to 2% of gross income.

Cost component items

In applying the principle of multicomponent subdivisions for cost analysis to sugarcane, the author (1977) has categorised no less than 55 subject areas for study, examined in detail 25 favourable combinations of conditions for cost estimations and required 35 subheadings to summarise the findings effectively and tabulated under 20 sets of conditions, recommendations related to areas of research and development considered to be important. Whilst this study was specifically for the use of sugarcane, the general principles apply to any agro-source with similar applications of criteria and constraints. The study was comprehensive but not necessarily exhaustive.

The author has preference for the following costing subdivisions but recognises that adjustments are permissible to accommodate other personal preferences:-

1) Development of land including net capital cost, drainage and irrigation development
2) Cultivation costs
3) Fertilisers and chemicals
4) Irrigation application
5) Harvesting
6) Management of plantations and farms
Subtotal A = total growing costs
7) Transportation of raw material from field to factory
8) Capital cost of factory
9) Maintenance
10) Labour and chemicals for factory operations
11) Management of factory
Subtotal B = total processing costs
C = total cost of ethanol ex-distillery
C = A + B

All agricultural operations have become highly mechanised with a few exceptions such as rubber tapping. But mechanisation has its price to pay - economic application demands continuity of use of equipment; this in turn requires a high standard of maintenance to be achieved and also that the items of equipment be used for as large a proportion of available time as possible. The seasonal nature of most agricultural operations reduces the time available for use in any one year, and anything which can be done to extend seasonal operations must be of benefit from this point of view. Agricultural machinery is generally not distinguished for the standard of maintenance achieved, perhaps more so when small scale units are employed. Observations made by the author have indicated that agricultural machinery is seldom used for more than 10-20% of its potential capacity.

An analysis of lost time due to mechanical failures for agricultural equipment in general reveals maintenance standards close to one-tenth of those achieved within the factory. Continuity of operation of other agricultural operations also has economic benefit, and optimisation is required at all stages.

THE SAGOPALM AS A POSSIBLE CROP

The sagopalm as grown in Malaysia would seem to have many advantages for the development of an alcohol fermentation industry. On the agricultural side it grows in low grade swampy country requiring the minimum of land preparation and agricultural attention. Growth conditions and yields have been well set out by Flach (1977) and have been used here as a basis for the calculation of potential ethanol yields.

Table 1. Annual Yield of Ethanol from Several Possible Crops

Plant crop	Carbohydrate Nature	Yield t/ha	Comment	Ethanol yield @ 88% recovery - kl/ha
Sago	starch	73	achievable	46
		25	cultivated	16
		10	semi-wild	6.3
		7	wild	4.4
Sugar-cane	sucrose	100	achievable	60
		28	good cultivation	17
		12	"average"	7
		8	poor but not uncommon	4.2
	sucrose and cellulose	150	achievable	78
		42	good	22
		20	"average"	9.6
		11	poor	5.6
Sugar-beet	sucrose	9	good	5.4
		4.5	average	2.7
Cassava	starch	13	achievable	8.2
		6	good	3.8
		4.5	average	2.8
Potato	starch	10	good	6.3

In Table 1 are listed the kind of yields which might be expected from sagopalm cropping and compared with sugarcane, sugarbeet, cassava and potato. It can be seen that when grown under cultivated conditions the sagopalm yields compare more than favourably with the very best of sugarcane.

The sagopalm would appear to have many advantages over the growing of sugarcane in the equatorial climate of Malaysia. It prefers swamp land requiring minimal preparation and cultivation.

Harvesting is developed more along the lines of timber logging and would seem to avoid many of the problems experienced with harvesting sugarcane. Present experience indicates no requirements for fertiliser, and natural rainfall can provide all of the water needed during the growing cycle. Harvesting can be conducted throughout the year, enabling processing costs to be kept lower than for seasonal crops.

There are however some disadvantages, but not of an insuperable character. The palms require 4-5 years before the starch production period commences. This might well be reduced by appropriate breeding as has been achieved in the case of oilpalms. Some attention might well be given to the technology involved in debarking and grinding.

The production of ethanol involves the use of energy for distillation and related processing, which means generating steam and associated generation of electricity for the plant requirements. Sugarcane has advantages in being able to use the fibrous component as a fuel for this purpose. It is possible that the sagopalm could do likewise since both bark and leaves could be considered as a potential source of fuel; this will be considered in more detail later.

The disposal of slops from the distillery poses a potential environmental problem which will call for a solution related to the particular situation of the sagopalm. One technique is to evaporate the slops and use the product as fertiliser. This is expensive in the consumption of energy and can increase steam requirements by as much as 50%; such a product is an expensive fertiliser and careful costing for disposal is required. Much depends on the size of the factory being operated. With the rising price of fuel oil the economics of employing the palm's own fibrous material for fuel becomes more and more attractive.

The present price of sagopalm delivered to the factory is around M$ 16/ton which is equivalent to M¢ 15/l for alcohol at 88% recovery. This represents a small enough return to the individuals concerned with production, harvesting and transport.

For development of a production facility of the size suggested, viz. 200 Ml/year, it is essential to maintain a

regularity of supply of raw material not only on a day-to-day
basis but year-in and year-out. This involves maintaining a
high standard of discipline on the part of the suppliers relating
to quality of raw material as well as to quantity; it is difficult
to envisage this happening if dependence is entirely on palms
grown "wild" and almost certainly predicates the need for systematic cultivation. It also implies the need for better mechanisation for harvesting and development of an efficient transport
arrangement. With these in mind it would not be difficult to
envisage the cost component of the raw material being reduced
by 50% to M¢ 7.5/l at the factory site.

One big advantage which the sagopalm seems to possess over
any other agricultural raw material is the ability to hold it
in the form of cut logs with only minimal deterioration of starch
content for periods reported to be as long as three months.
This makes it possible to provide a very useful insurance against
the vagaries of weather and other factors which can interfere
with a regular supply to the factory.

It is difficult to suggest a cost estimate for the erection
of a factory and associated infrastructure, but it is thought
that a figure of the order of M$ 85 million should be adequate
for this purpose for preliminary cost estimating. If this sum
be amortised over a period of 20 years at 20%, it would represent
an annual charge of 20.45% or M¢ 8.7/l; to this should be added
a maintenance cost equivalent to 5% on capital or M¢ 2.1/l.

The cost of operating the distillery will be related very
largely to the degree of automation incorporated. The capital
charge cited above is believed to be adequate to provide a high
degree of automation for which labour requirements would be a
minimum. Such an arrangement would be in keeping with the primary
predication that the standards of operation should be those
established by the oil industry for which a refinery would be
a suitable comparison.

For a labour force of 100 men at an average cost of M$ 0.6/l,
should be added say M¢ 0.4/l for management costs; chemicals
and laboratory control of a high standard would be required and
for which a charge of M¢ 1.0/l should be allowed.

The operation of the distillery requires fuel to generate steam for the distillation and electrical energy to be used for operating pumps, compressors and other mechanical equipment; there will also be slops as a waste material from the stills. The energy requirement for all operations including the processing of slops will vary between 3.5 and 5.5 kg steam per litre of alcohol produced. If we take the higher figure and initially cost it for generation from fuel oil, it will give an indication of the value which might be placed on alternative types of fuel and the costs which might be economically incurred for their development.

The present price of fuel oil in Malaysia is around M$ 300/ tonne, which is normally related to the price of crude oil and maintained slightly below; if the price of fuel oil exceeds the price of crude then users of fuel oil will prefer to use crude. A figure of M$ 300/t for fuel oil corresponds to US$ 18/bbl and relates to a crude oil figure around US$ 20. With oil able to generate 16 kg of steam per kg of oil, it would require 0.34 kg of oil to produce a litre of alcohol at a cost of M¢ 30/kg that would represent M¢ 10.3/l of alcohol. As this would come very close to an expenditure of M$ 21 million per annum, it can be seen that there is ample scope for serious costing for alternative fuel sources as well as for economy of steam usage in processing.

Table 2 represents a summary of itemised costs of producing industrial alcohol from sagopalm at three scales of operation. The scale of operations at 200 Ml alcohol/annum would require 12,500 ha of cultivated sagopalm at 25 t starch per ha. Price of sagopalm delivered to factory = M$ 16/t @ 18.5% starch, with 91% recovery of starch and 88% recovery of theoretical alcohol production from starch (theoretical 720 l alcohol per t starch).

The cost of the largest scale of operation compares more than favourably with the estimated costs of alcohol produced from sugarcane, which might vary from around M¢ 75 to perhaps as low as M¢ 25 under ideal conditions, which are unlikely to be realised in an equatorial climatic zone and for which a median figure of M¢ 50 might represent reasonably comparable constraints.

It may well be preferred to commence operations on a smaller

Table 2. Estimated Costs of Anhydrous Ethanol Produced from the Sagopalm at Three Scales of Operation

Scale of operations (Ml alcohol/yr)	200	20	2
Estimated capital cost (M$ million)	85	22	5.4
Item		M¢/l	
Raw material	15.0	15.0	15.0
Capital charges for factory (amortisation for 20 yrs @ 20%)	8.7	22.5	55.2
Maintenance @ 5% on capital	2.1	5.5	13.5
Labour costs	0.6	1.5	3.8
Chemicals and control	1.0	1.2	1.5
Management	0.4	1.0	2.5
Fuel	10.3	12.0	15.0
Total cost ex-distillery	38.1	58.7	106.5

scale, for which higher costs could be anticipated. For a scale of operations of the order of one-tenth of those suggested, i.e. for 20 Ml/annum, the costs might be expected to increase to 60¢/l ex-distillery, and for a very small factory producing 2 Ml/annum the cost would be close to M$ 1.10/l.

There might well be merit in building an initial plant to process 20 Ml/annum for which a capital requirement of the order of M$ 22 million could be anticipated and which could be viewed as a plant of an experimental character. The oil industry has invariably maintained that large capacity is necessary for costs to be contained in what is essentially a capital-intensive industry, and it can be seen that a similar comment could be applied to an agro-ethanol undertaking reflecting corresponding cost benefits from any increase in the scale of operation. The 2 Ml/annum plant would still require something of the order of M$ 5-6 million for the distillery and infrastructure.

The tops and leaves could be used as fuel just to generate electricity for the national grid at a cost no more than that of hydropower. Associated with a 200 Ml distillery could be operated a 50 MW power station able of itself to substitute for the purchase of fuel oil to the value of something like M$ 50

million providing a further credit equivalent to a net of
10-12¢/1.

The Energy Balance

It has been well pointed out by Greenfield and Nicklin (1979)
that there is not much point in producing a fuel if the energy
required for its production is greater than the net energy of
the product. This can happen very easily if technology is not
established at an appropriate standard, and this comment applies
to agro-ethanol. If fuel oil must be burned to provide steam
for the ethanol distillation columns a very careful cost-
accounting is necessary. Different technologists use different
bases for calculating the energy balance. Four are considered
here.

Anhydrous ethanol itself has a net thermal value of 21.3 Mj/l
for which the energy required in processing and distillation is
13.5 (± 3), which represents a net energy gain of 58 (± 13)%
with respect to the energy consumed. From this must be subs-
tracted the energy required for cultivation, harvesting, transportation and fertilising of the crop, which in the case of the
sagopalm would probably not amount to more than 2(± 1)%. The
disposal of slops will also consume additional energy which
could well be as much as 40%, that would make a substantial
inroad upon the net energy value still showing a net gain but
of uncertain magnitude, which could be quite small if thermal
efficiency in the distillery is not maintained at a high level.

Another way of looking at the energy balance is to consider
first the energy value of the sagopalm itself as a fuel source
and compare this with the energy value of the product with its
upgraded energy content. Three views are possible on this basis,
one of which would refer to the energy value of the debarked
trunk, the second to that of the trunk with bark, and the third
of the whole palm.

If we use the analytical data of Flach, then at 80% moisture
content for the debarked trunk we are dealing with a fuel of
very low energy value indeed and comparable to that of peat in
its natural state. A net thermal value of 0.78 Mj/kg of trunk

would result in the production of fuel with a net thermal value
of 2.27 Mj, but for which would be required 1.2 for processing
and distillation plus a component for slops disposal, which
could go as high as 0.5. On this basis it can be seen that
there would be a net loss of energy of 27%; this could be just
reversed to a net gain of 11% if the thermal efficiency of the
processing and distillation be increased to maximum achievable
figures.

Flach quotes a figure of 0.1 kg of bark per kg of debarked
trunk but does not give any value for the moisture content of
the bark. Assuming that this is 20%, then the fuel value of
the bark works out at just about the right figure to provide
the processing requirements for the starch in the portion of
trunk from which the bark was removed; there could even be a
small surplus if good energy efficiencies are achieved. On this
basis there would be a net energy gain of 190% for which a
reduction of M⌀ 10-15/1 would be appropriate confirmatory evidence. There would be more difficulty in achieving the full
energy balance for the smallest scale of operation owing to
greater difficulties in obtaining high thermal efficiencies
with small scale equipment, but a reduction of M⌀ 10 of the
possible M⌀ 15/1 should be achievable.

The calculation of the fuel value of the whole sagopalm
is fraught with greater difficulties owing to the lack of factual
data concerning the composition of the whole palm. It is evident
however that the leaves and top sections of the palm would
represent a source material having significant thermal value
in excess of that required for processing; if it is just wasted
then the product would show a significant thermal deficiency.
Any value as humus return to the soil would need to be carefully
evaluated. The possibility of converting its cellulosic component
to ethanol and thereby increasing the net yield per hectare
could have economic advantage but would require a higher standard
of technological expertise. Alternatively it might just be
collected and burned to generate steam and electricity of worthwhile economic benefit as has already been mentioned.

Other Operating Considerations

To achieve maximum efficiency involves both a high standard of chemistry and a high standard of engineering. The chemistry involves hydrolysis and fermentation which require well informed biochemical control in order to maximise conversion of carbohydrate to ethanol and to provide the most favourable conditions for recovery.

The concentration at which the biochemical reactions are operated has an important bearing on cost. The reactions do not favour high concentrations, but every additional litre of water employed to provide a more favourable biochemical environment involves the subsequent cost of separation of the ethanol from the more dilute conditions. The efficient recovery and recirculation of yeast is a necessary achievement.

Distillation technology needs to be the best of modern development employing multiple effect principles for feed pre-heating as well as in the distillation operation itself. Ethanol and water form a constant boiling point mixture at a concentration around 95% ethanol which is impossible to separate by simple distillation. However, special dehydrating distillation techniques are now well developed and there should be no technical problem in producing anhydrous ethanol after an extra stage of distillation. The added value of the anhydrous product is generally considered to justify the additional cost, except perhaps in small scale operations.

Cooling of fermentation vessels contributes an important component to the process energy balance. Carbon dioxide is a significant byproduct of fermentation and technology must be of a high enough standard to ensure thorough stripping from all ethanol. Occasionally it may have value as a source of the refrigerant "dry-ice" but more commonly it is vented to the atmosphere. In order to protect the community from undesirable social consequences of large scale ethanol production, it is necessary to denature the product, an exercise which in itself requires special study and concern.

Modern technology now makes possible almost complete automation of all unit operations involved with supervisory labour requirements, enabling them to be reduced to a very small component.

Future programming

Forecasting the future is an invidious occupation; nevertheless when an industrial operation is of the magnitude discussed here, it becomes necessary to anticipate a sustained period of uninterrupted production with raw material supplies and market conditions keeping a satisfactory relationship.

To operate a plant at 200 Ml/annum, i.e. rated at 550 kl/day, relates to a moderate scale and in fact is small in terms of petroleum refineries. This would provide for approximately 10% of Malaysia's petrol needs and would have to be distributed in the form of a blended motor-spirit. To supply such an amount of ethanol would require the continuous cropping of 12,500 ha of cultivated sagopalms or 50,000 ha of wild palms.

It can be seen from Table 2 that costs for a distillery operating at 20 Ml/annum would still be containable within the present world price of oil, and there could be advantages from the experimental point of view in commencing at this scale. To operate at only 2 Ml/annum would incur high relative costs, and one means of containing these is to produce only 95% ethanol which is suitable for blending with petrol in concentrations up to 5%.

Looking at the prospect in the other direction, that of expansion, it has been estimated that there are something of the order of 2 million ha of swampland available in Malaysia and which if developed to a productivity level of 16 kl/ha would represent a projected 32,000 Ml per annum, which far exceeds the likely usage even for 100% replacement of petrol (Australia's current consumption is around one third of this figure). Nevertheless, if costs are contained within those set out in Table 2, Malaysia could become an exporting country operating on a continuous basis and on a scale even beyond that of the present oil wells, which could be exhausted by the time the 2 million hectares are developed.

Establishing and maintaining low cost conditions in an

agro-ethanol industry have been discussed elsewhere by the author (1979) and there would seem to be no insuperable technological barriers to ensure such prospects. A final emphasis is however deemed desirable, that the investment of money in research and development is just as important as it even has been in the fossil-oil industry.

REFERENCES

FLACH, M. 1977. Yield potential of the sagopalm, Metroxylon sagu, and its realisation, Sago-76: Papers of the First International Sago Symposium, ed. K. Tan; Kuala Lumpur: 157.

GREENFIELD, P.P., and D.J. Nicklin 1979. Synthetic fuels – the interaction between energy accounting and cost accounting, Australian Institution of Engineers National Conference, 1979, Newcastle (Australia): 31.

KELLY, F.H.C. 1977. A Feasibility Study of the Production of Ethanol from Sugar Cane; Dep. of Chemical Engineer. monograph, University of Queensland, Brisbane: 218 pp.

KELLY, F.H.C. 1979. Cost control factors in the production of ethanol from sugarcane. UNIDO Workshop on Fermentation Alcohol, 1979, Vienna; ID/WG/293/15.

THE USE OF INDUSTRIAL ENZYMES IN THE CONVERSION OF STARCH
TO FUEL ALCOHOL

J.H. BAKER

AGROFUEL

Due to the continually increasing price of oil, many countries are being forced to consider alternative liquid fuels. It
has already been proven that up to 20% ethanol, produced by
fermentation of locally available raw materials, can be mixed
with gasoline without requiring any changes to be made to automobile engines. Such a system should reduce a country's oil
importation costs and supply jobs for many thousands of people.
The use of fuel alcohol is already well established in Brazil,
and many other countries are either currently introducing the
concept or proposing its introduction in years to come.

Raw materials for fuel alcohol production include molasses,
sugarcane juice and starch. Molasses and sugarcane juice are
already in a fermentable form, hence making their conversion into
alcohol a simple process. However, the problem is that most
countries do not have sufficient year-round supplies of these
raw materials to produce the enormous quantities of alcohol that
would be required to support a national programme. This is where
starch comes in as it can be used to supplement the deficiencies
of the sugarcane derived fermentables.

In many Southeast Asian countries starch-containing plants
are readily available or can be easily grown. However, starch
is not fermentable as such and hence must be first converted
into a fermentable form. This conversion is done most economically by the use of industrial enzymes known as amylases. The
starch referred to throughout is in a pure or relatively pure
form; process parameters mentioned may therefore vary when ground
starch containing plant material is used as a raw material.

*W.R. Stanton and M. Flach (eds.), SAGO. The Equatorial Swamp as a Natural
Resource. Proceedings of the Second International Sago Symposium. All rights reserved.
Copyright © 1980 Martinus Nijhoff Publishers, The Hague/Boston/London.*

THE CHEMICAL AND PHYSICAL NATURE OF STARCH

It appears that starch from any source can be used as a substrate for enzymatic conversion into fermentable sugars. Common starches which have been investigated by NOVO INDUSTRI A/S (the world's largest producers of industrial enzymes) include potato, corn, wheat, tapioca (cassava), rice and sago.

Structurally, starch is made up of glucose units linked together by alpha 1,4 and alpha 1,6 linkages. Linear starch chains which consist of only alpha 1,4 linkages are known as amylose molecules; 20% of most starches are in this form. Branched starch chains are known as amylopectin molecules; this branching is due to the presence of the alpha 1,6 linkages.

In its native state starch consists of insoluble microscopic granules which must be first disrupted before enzymatic attack is possible. The disruption process is known as gelatinisation and is initiated by exposure of a starch slurry to high temperatures (above $60^{o}C$); during such a heat treatment the granules swell and eventually rupture, dispersing the starch molecules into solution. Gelatinisation temperature varies with different starches. It appears however that wheat, tapioca, potato and sago are the easiest of the common starches to gelatinise in terms of temperature requirements.

Enzymatic Starch Conversion

For conversion of a starch molecule into glucose units two kinds of enzymes are required.

Bacterial alpha amylase

This enzyme randomly hydrolyses the alpha 1,4 links of the starch molecule to produce somewhat shorter glucose polymers known as dextrins; such a process may be referred to as liquefaction or dextrinisation. There are two types of industrial bacterial alpha amylases available on the market today. The first type, BAN^{R}, which can be called a traditional bacterial alpha amylase, is produced by fermentation of a selected strain of Bacillus subtilis. The second type, TERMAMYL^{R}, is an extremely heat-stable alpha amylase and is produced by fermentation of a selected strain of B. licheniformis.

TERMAMYL[R] is less dependent on calcium ions for its stability than BAN[R], e.g. 50 ppm vs 150 ppm calcium respectively, and the higher heat stability of TERMAMYL[R] is of benefit in certain continuous starch conversion processes where only one addition of the enzyme is desired. When starches with a high gelatinisation temperature are being processed, e.g. rice, corn, the use of TERMAMYL is also advantageous as higher retention temperatures can be maintained without any significant loss of enzyme activity.

Amyloglucosidase (Gluco amylase)

Industrially, amyloglucosidases are most commonly produced by fermentation of a selected strain of the fungus Aspergillus niger. This enzyme, SAN[R], converts the dextrin molecules into glucose units by hydrolysing both the alpha 1,4 and alpha 1,6 linkages. Such a process is known as saccharification.

GUIDELINES FOR CONSTRUCTING A CONTINUOUS STARCH CONVERSION PROCESS

As a result of practical experience and a number of laboratory and pilot plant studies, NOVO INDUSTRI A/S can offer the following guidelines for the continuous conversion of starch into fermentable sugars suitable for alcohol production.

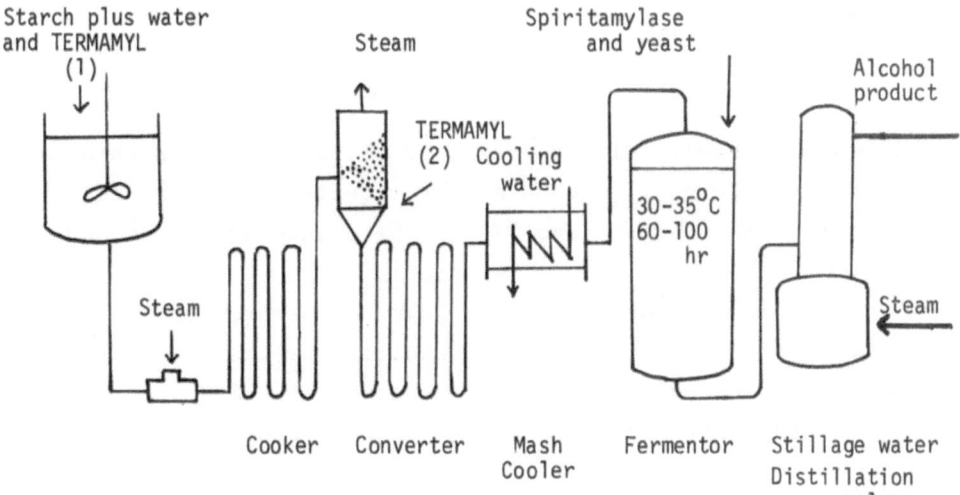

Continuous Starch Conversion Process

Preparation of the Starch Slurry

The initial stage in the production of alcohol from starch
is the preparation of a slurry containing the correct starch dry
solids (DS) concentration. Several factors must first be considered before preparing the starch slurry.

(a) Alcohol yield: the maximum theoretical yield of alcohol
which can be obtained from starch is 715 litres of alcohol per
tonne of starch DS (Appendix). In a practical situation however
it is found that the maximum yield of alcohol obtainable is around
660 litres per tonne starch DS. This is due to the yeast biomass
production as well as the yeasts metabolic processes producing
other compounds apart from ethanol.

(b) Alcohol percentage present in final fermented liquor:
during fermentation the yeast will produce a maximum 7-9% v/v
alcohol due to the inhibitory effect of alcohol on yeast
metabolism.

Based on the above facts it is recommended that the initial
fermentation liquor contain 15-20% DS based on starch. When
saccharified, this percentage of starch DS will give the required
concentration of glucose for maximum alcohol production. The
correct glucose concentration may be arrived at by one of two ways.

(i) The simplest way of course is to prepare a 15-20% starch
DS slurry in the first place. The disadvantages of doing it this
way are:-

(a) Termamyl stability: the lower the starch DS the less
stable is Termamyl. With a starch DS of 15-20% it has been found
that a double Termamyl dosage may be required: 1/3 of the total
dosage at first addition and 2/3 of the dosage at the second
addition.

(b) Equipment capacity: the lower the starch DS the larger
are the actual volumes of liquid requiring processing. If the
volumes could be reduced by using a higher starch DS slurry the
equipment capacity could be reduced, hence resulting in a nett
saving on construction costs.

(ii) An alternative process suggestion involves the initial
preparation of a 35-40% starch DS slurry. At this DS only one
enzyme (Termamyl) dosage is necessary (enzyme only dosed to the

slurry tank). Due to the higher DS the equipment capacity can
be reduced as compared with process (i) above. In order to
arrive ultimately at the final 15-20% DS based on starch, the
slurry can be diluted after liquefaction by the addition of the
correct amount of water; dilution could take place either in,
or just prior to, the fermentation tank. Such a system would
also have the advantage of cooling the liquefied starch slurry
close to the required fermentation temperature, depending on
temperature of dilution water used.

Once a starch slurry of desired DS percentage is prepared,
the pH is adjusted, if necessary, to 6.0-6.5; this is the optimum
operating pH for Termamyl. In order to increase the heat stability of the enzyme, at least 50 ppm of calcium should be present,
which is normally however present in the water used to prepare
the slurry. The enzyme is also added at this stage to the slurry
tank and the starch solution is ready to be gelatinised and
hydrolysed.

Traditional gelatinisation processes have usually involved
pressure cooking the starch solution at 130-160°C. Novo's experience, however, suggests that any common starch can be completely
gelatinised if the slurry is passed through a jet cooker (steam
injection heater) and held for 5 minutes at 103-107°C; the jet
cooking device instantaneously heats up the starch slurry to the
desired operating temperature, which is considerably lower than
that used in the traditional pressure cooking process. This
reduced operating temperature will result in significant savings
on steam cost.

The injection of steam into the slurry also exerts a high
shear; this shear effect supplements the action of the enzyme by
partially splitting up starch molecules. After a five-minute
period in a series of retention tubes, the slurry is flash-cooled
to 80-90°C and held for another 20 minutes at this temperature in
a second series of retention tubes; if a 15-20% starch DS solution
is being processed, a second Termamyl dosage may be required here.

Finally, the slurry is cooled to fermentation temperature
(30-35°C) and added directly into the fermentation tank together
with the yeast and SAN (saccharification enzyme). The dextrins

are thereby converted to glucose in the fermentor; simultaneously the yeast converts the glucose produced to ethanol and CO_2.

Fermentation Time

The fermentation time for such a process is of the order of 60-100 hours, after which time the fermented liquor is distilled to produce pure alcohol for fuel addition. Traditional alcohol production from molasses or sugarcane juice takes in the order of 24-36 hours.

The reason for the great difference in fermentation time (compared with the starch process) is that in molasses and cane juice all the sugar present is already in a fermentable form. However, with starch the conversion rate into glucose decreases towards the end of the fermentation when the more difficult alpha 1,6 links and very short chains are hydrolysed.

The fermentation time can be reduced by the addition of some fungal alpha amylase to the fermentor together with the SAN; increasing the fermentation temperature; addition of a higher initial inoculum of yeast; and increasing the dosage of SAN. These factors will however not reduce the fermentation time to 24 hours, but 48 hours is not uncommon, and 36 hours has been used in some cases.

FUTURE DEVELOPMENTS IN STARCH CONVERSION FOR ALCOHOL PRODUCTION

Fuel alcohol is still in its infancy and much research is still needed in order to streamline the processes of fermentation and distillation. It is believed, however, that we will soon see the introduction of continuous fermentation systems to fuel alcohol plants. The main advantage of continuous fermentation as compared with the traditional batch process is that the total fermentation volume can be substantially reduced, by about a factor of 10.

A change from conventional fermentation to continuous fermentation will require modification of the saccharification process herein described. With continuous fermentation the saccharification will then require optimisation to give maximum glucose yields. The starch liquefaction process which precedes saccharification and fermentation will also require slight modifications in terms of retention times.

As the fuel alcohol concept becomes more widespread, technological breakthroughs are bound to reduce the overall costs, hence making fuel alcohol an even more viable alternative to imported petroleum.

Appendix: Theoretical Yield Calculations for the Conversion of Starch into Ethanol

The stoichiometry of the conversion is as follows:

$$\text{Starch polymer} \xrightarrow[\text{Enzymatic hydrolysis}]{+H_2O} n \; \underset{\text{Glucose}}{C_6H_{12}O_6}$$

$$C_6H_{12}O_6 \xrightarrow[\text{Yeast fermentation}]{} \begin{array}{c} 2\; C_2H_5OH \\ \text{Ethanol} \\ + \\ 2\; CO_2 \end{array}$$

It can be seen that the hydrolysis of starch involves a nett weight gain due to the addition of one water molecule per glucose molecule released.

Theoretically therefore:

1) 1,000 kg starch give 1,111 kg glucose;
2) From stoichiometric relationships 180 g glucose is converted to 92 g of ethanol plus 88 g of carbon dioxide;
3) The specific density of ethanol is 0.79, hence 92 g of ethanol fills a volume of 116 ml;
4) From combining (1), (2) and (3) we obtain the following:

$$1{,}111 \text{ kg} \times \frac{92}{180} \times \frac{116}{92}$$

$$1{,}111 \text{ kg} \times \frac{116}{180} = 715.9 \text{ litres}$$

Theoretically, 715.9 litres of ethanol can be produced from a tonne of starch.

BIOCONVERSION OF STARCH INTO PROTEIN

M. RAIMBAULT

STARCHY SUBSTRATES

Starchy materials, more specifically cassava and sago in
tropical regions, are of great interest owing to both their
high productivity per unit area and excellent rate of conversion
to microbial protein by a large number of fast growing microorganisms.

For food production, the question arises if it is better
to cultivate a protein-rich plant of relatively low productivity
or a highly productive starch-rich plant, being settled that
starch can be transformed into protein at an average rate of
25%. Table 1 clearly demonstrates that it is more promising,
from the aspect of food supply, to cultivate starchy plants for
supplying both calorie and protein.

LIQUID FERMENTATION

The problem is to ascertain which are the processes available
for transforming starch into protein by microorganisms. Much
work was carried out recently in this field, but available
processes are essentially based on liquid fermentation technology.
They can be classified into three groups:

1) Two-step processes - starch hydrolysis using amylolytic
enzymes, amylase or amyloglucosidase, then yeast culture
on hydrolysate; these processes require sterilisation
and aseptic conditions.

2) One-step processes with two organisms - one amylolytic
fungus and one yeast, known as the Symba Process;
sterilisation and aseptic conditions are required too.

3) Direct growth of filamentous fungus to produce biomass;
some processes are described that do not require

Table 1. Productivity of Some Foodcrops

Crop	Yield t/ha	Protein t/ha
Protein-rich plants		
Soybean	1.8	0.6
Sunflower	2.5	0.6
Horse bean	3.2	0.9
Pea	3.0	0.75
Rapeseed	3.0	0.7
Cereals		
Corn	6.0	1.9
Wheat	5.0	1.7
Starch plants		
Cassava [a]	12.0	3.0
Sago [b]	14.5	3.5

[a] Cassava potentiality is calculated on the basis of 40 t/ha of fresh roots

[b] Sago production is calculated on the basis of extensive cultivation, with spacing of 6 m between palms and a yield of 500 kg of crude meal per trunk after 8 years.

sterilisation and aseptic conditions. The most interesting feature is the facility of harvesting the biomass; on the other hand, filamentous fungi increase the viscosity of the liquid.

Technically, all these processes exist and are quite feasible. They provide biomass with high content in protein of excellent nutritional quality. However the protein cost is not competitive with conventional protein production because of the high capital investment and energy demand, and the low price of conventional protein; but this last point could be changed in the future if the protein shortage happens, or if palm oil competes with soya oil.

The liquid processes are mainly used in starch factories when it is important to treat the effluent in order to avoid too much pollution of the environment; in this case, the price of the substrate is negative.

SOLID FERMENTATION

Another promising technology is solid fermentation, so called in contrast with liquid fermentation. Solid fermentation in Southeast Asia is exemplified by many traditional fermentations, e.g. tempeh, ragi, koji. However, these fermentations do not increase the protein content, but rather improve acceptability, digestibility or flavour of the food.

Knowledge of solid fermentation is not as advanced as the liquid fermentation. Nevertheless, Hesseltine for tempeh processing and Stanton for cassava fermentation carried out very interesting and important researches in this field of solid fermentation. Since 1974 I have investigated at ORSTOM the possibility of low technology processes to enrich the protein content in starchy materials by direct growth of fungi in solid fermentation, in order to obtain a product containing not necessarily a very high protein content but enough to make it usable for animal feeding, viz. 15-20%.

At first, I fitted up a solid fermentation method of culture for studying microbial, physiological and biochemical aspects of the growth of fungi in such a solid fermentation. The principle of the technique is based on the homogeneous distribution of spores and mineral salts in the mass of the starchy material put in suitable form. The preparation of a porous granulated material with adequate pH, temperature and moisture content is essential to ensure good aeration and fast growth of mycelium all in the mass.

The coarsely ground raw material with 30-35% moisture is maintained at 70-80°C for 10 minutes by gentle steaming in order to gelatinise starch granules; after cooling to 40°C, this steamed substrate is mixed with water containing the inoculum of spores and mineral salts to 55% moisture content. For laboratory purposes we adopted a very simple incubator (Figure 1). This method has already been worked out with a variety of starchy materials, namely cassava, whole potatoe, potato waste from industrial fecula works and banana refuse. We have not tried sago yet, but I think it would be interesting to test this starchy material. The results are reported in Table 2.

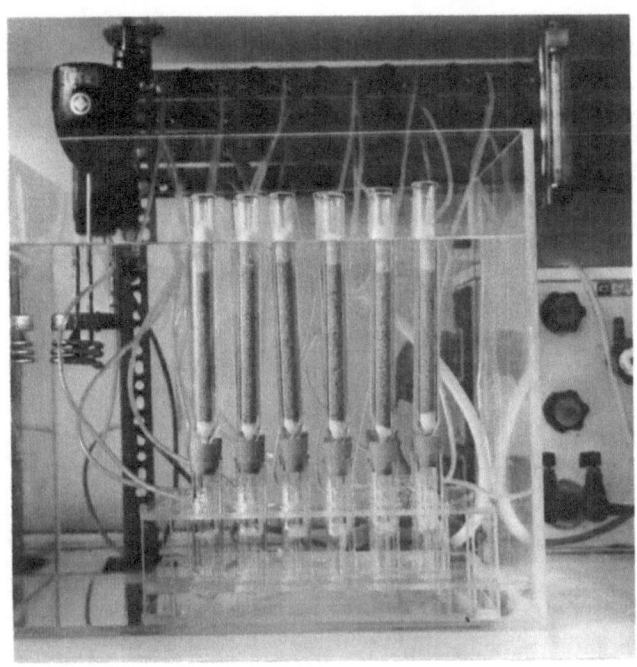

Figure 1.   Laboratory Design for Solid State Fermentation

Experiments were performed with a selected strain of
Aspergillus niger, having high amylolytic activity and suitable
amino-acid composition.   Many amylolytic fungi, particularly
among strains used in Asian traditional fermentations for human
consumption, were successfully tested by this technique with
comparable results.   This method does not require aseptic

Table 2.   Protein Enrichment from Several Starchy Substrates

| Substrate | Initial Product | | Final Product | |
| | Protein | Carbohydrate | Protein | Carbohydrate |
		g/100 g DM		
Cassava	2.5	90	18	30
Banana	6.4	80	20	25
Banana waste	6.5	72	17	33
Potato	5	90	20	35
Potato waste	5	65	18	28

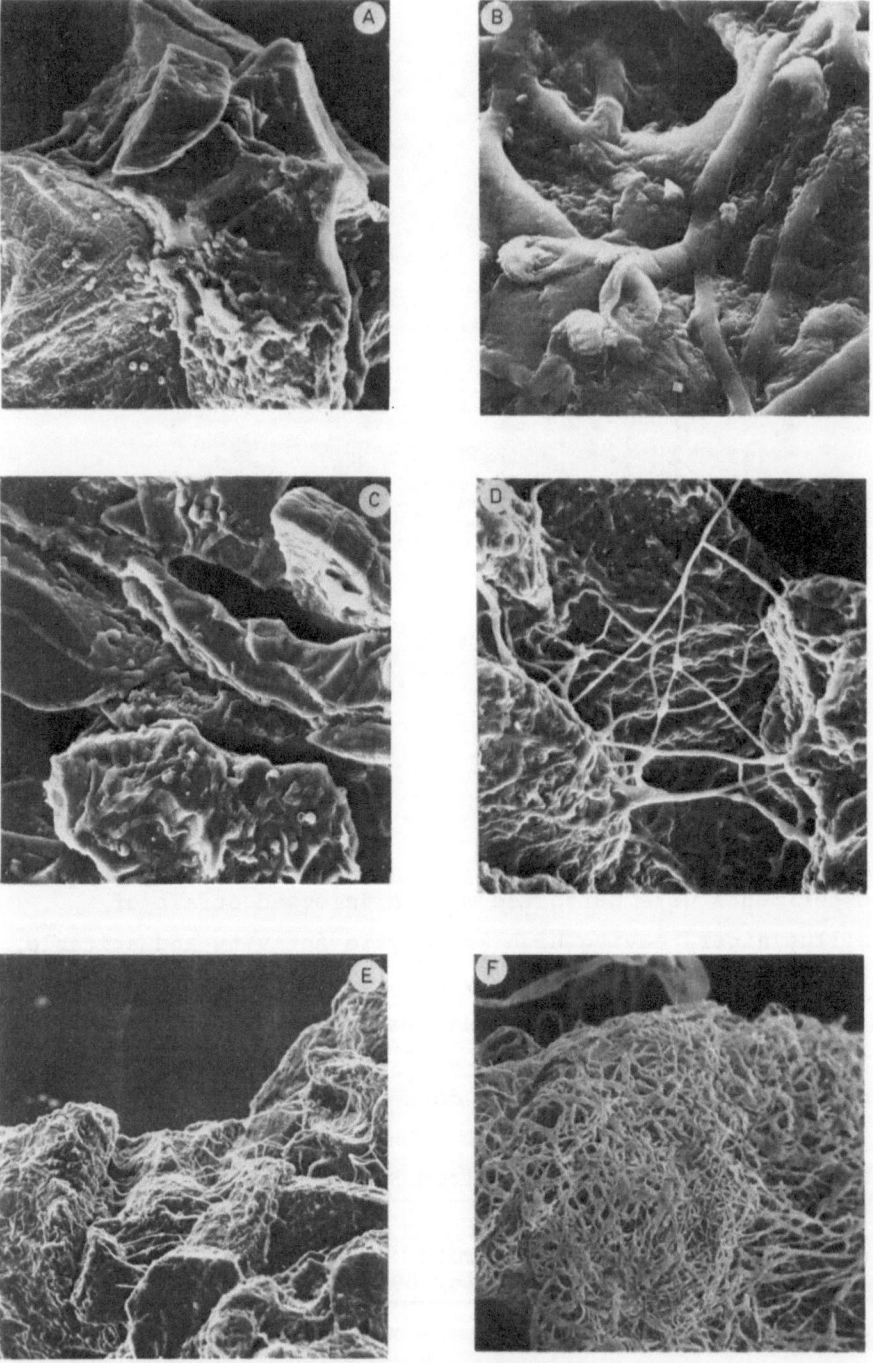

Figure 2.   Growth of the Fungus in Starchy Substrate during
           Solid State Fermentation

A : t_o;   B : t = 8 h;   C : t = 12 h;   D : t = 16 h;
           E : t = 20 h;   F : t = 24 h.

conditions, selective growth of the fungus resulting from acidic pH, low moisture content and heavy spore inoculation.

I want to show you a series of photographs demonstrating the development of the fungus inside the starch substrate, with the aid of scanning electron microscopy (Figure 2).

From laboratory experimentation, equipment was designed for the solid state enrichment process at the pilot scale (Figure 3).

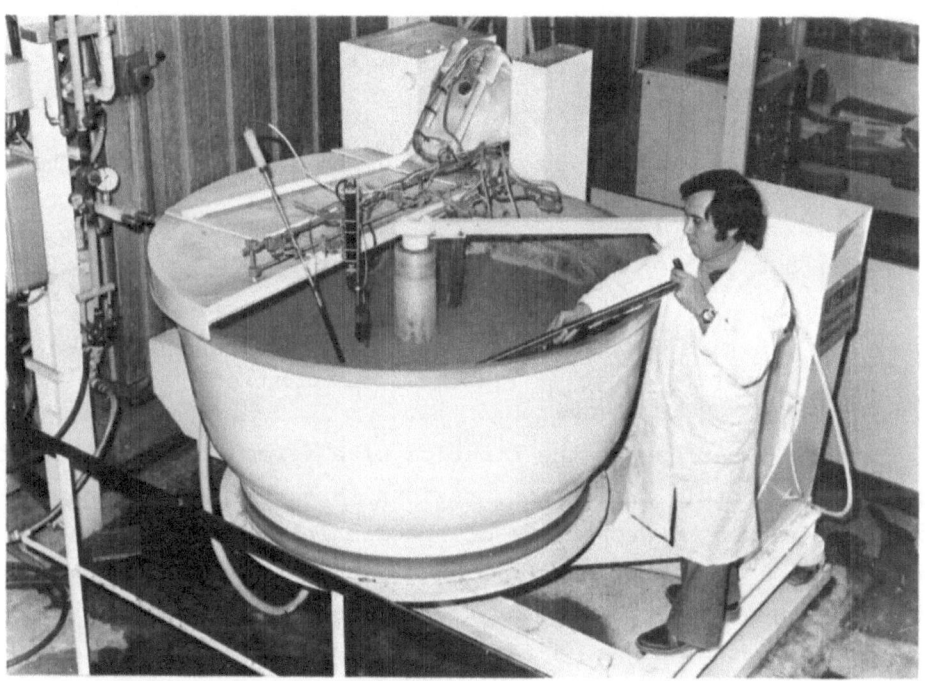

Figure 3. Pilot Scale Fermentor for Protein Enrichment by Solid Culture (200 kg DM capacity)

All the operations are conducted in a commercial bread-making blender modified for the purpose. Steaming and aeration are performed by passing steam or air through the perforated bottom of the tank. A simple control system using conventional probes was designed to keep suitable pH, moisture and temperature; this control system is activated by the temperature sensor as soon as temperature reaches the set point.

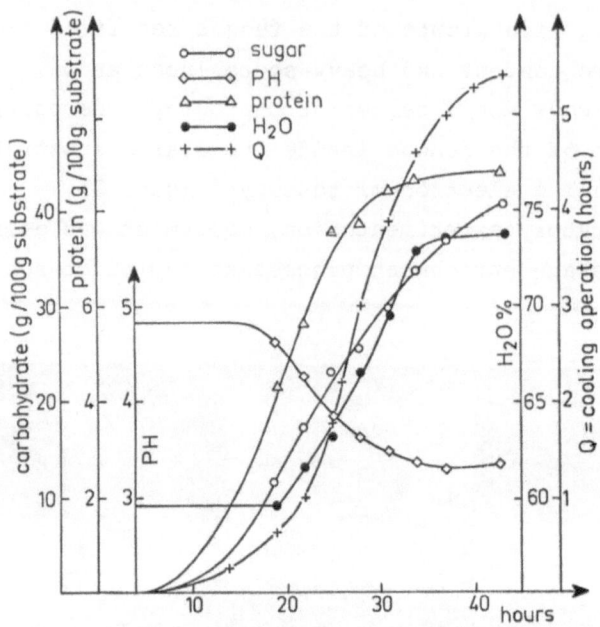

Figure 4.   Kinetics Evolution during Solid State Fermentation
of Potato Waste in Pilot Experimentation.

Figure 4 represents the kinetics of a fermentation on potato
waste, with protein production, residual sugars and moisture.
The curve marked by crosses is of special interest; it shows
that during a fermentation of 30 hours, the mechanical mixing
and spraying had to operate for five hours only, indicating a
low expenditure of power, a fact of obvious importance in regard
to production cost by solid state fermentation and to its economical feasibility at a low-level unit operation.

Presently a fermentor unit of one cubic metre, capacity
250 kg dry matter, is tested in a potato factory for protein
enrichment of wastes. This equipment will be used for large
scale nutritional and toxicological testing on target animals
(pig, poultry) and for the determination of the actual investment
and operation costs. It is intended that the experimentation
will be extended to tropical countries, in order to adapt the
procedure to local climatic and agroeconomic conditions and
also to different potential starchy substrates like cassava,
banana and sago for animal feeding purposes.

VALUE OF STARCH FERMENTATION

When we compare biomass production by liquid fermentation and protein enrichment by solid fermentation, we are surprised that the rate of conversion of starch into protein and the growth rate of the fungus are quite similar; the solid technology being very simple would have to be more competitive. Finally, starch is a very promising substrate for the future and starchy products will certainly become essential from the aspect of solar energy storage and fuel or food energy supply.

ORSTOM: Office de la Réchèrche Scientifique et Technique Outre-Mer (Overseas Scientific and Technical Research Bureau)

THE ROLE OF LACTOBACILLI IN STARCH ASSISTED FERMENTATION

YEOH Quee Lan

MICROORGANISMS

Microorganisms have long been employed by man for the preservation of raw materials for food. Especially in the tropics
where the warm ambient conditions favour the rapid growth of
microorganisms, man has to wage a constant battle to preserve
his food supply against spoilage organisms. On the other hand,
by manipulating the conditions of growth, it is possible to eliminate spoilage organisms and encourage the proliferation of
the desired microflora. Food fermentation is not without its
hazards and unless careful control of the fermentation conditions
are maintained, problems of microbial food poisoning may result
(Stanton and Yeoh 1978).

Lactic acid bacteria

Only a few species of microorganisms are important in food
fermentations. The lactic acid bacteria are a group which plays
a major role in many food fermentations. They may be found in
the commensal flora of many habitats (Table 1), but their common
characteristic is that they are able to produce lactic acid as
the major end product.

The lactic acid bacteria are generally recognised as consist-
ing of "Gram positive, non-sporing, carbohydrate fermenting lactic
acid producers, acid tolerant, of non-aerobic habit and catalase
negative, typically they are non-motile and do not reduce nitrate"
(Ingram 1973).

Physiologically, the lactic acid bacteria can be divided into
two subgroups defined by the products of glucose fermentation.
All strains which produce 1.8 moles of lactic acid per mole of

W.R. Stanton and M. Flach (eds.), SAGO. The Equatorial Swamp as a Natural
Resource. Proceedings of the Second International Sago Symposium. All rights reserved.
Copyright © 1980 Martinus Nijhoff Publishers, The Hague/Boston/London.

Table 1. Some Common Sources of Lactic Acid Bacteria

Sources	Desirability
Milk products	
Sour milk	no
Butter	yes
Cheese	yes
Yoghurt	yes
Meat and meat products	
Fresh meat	no
Fermented sausages	yes
Fermenting vegetables	
Cucumbers	yes
Sauerkraut	yes
Fodder silage	yes
Fermented beverages	
Beer	no
Wines	yes
Ciders	yes
Sake	yes
Man and animals	
Mouth	yes
Alimentary tract	yes
Blood stream	no
Lungs	no

Source: Adapted from Carr 1973

glucose with small amounts of acetic acid, ethanol and carbon dioxide are classified as the homofermenters. On the other hand, those that ferment hexoses with the production of less than 1.8 moles of lactic acid per mole of glucose, and in addition fairly large amounts of ethanol, acetate, glycerol, mannitol and carbon dioxide, are termed heterofermenters (Doelle 1969). The taxonomic subdivisions of the lactic acid bacteria are given in Table 2.

Different kinds of lactic acid, D, L, DL or mixtures of these acids are produced in each reaction and are characteristic for each genus or subgenus. Although all lactic acid bacteria can produce lactic acid, their acid tolerance is variable (Ingram and Luthi 1961).

As a group the lactic acid bacteria are fastidous in their nutrient requirements due to their extremely limited synthetic

Table 2.   Taxonomic Subdivisions of the Lactic Acid Bacteria

Cell shape and arrangement	Mode of glucose fermentation	Genus
Spheres in chains	Homofermentative	Streptococcus
Spheres in chains	Heterofermentative	Leuconostoc
Spheres in tetrads	Homofermentative	Pediococcus
Rods	Homofermentative	Lactobacillus
Rods	Heterofermentative	Lactobacillus

Adapted from Stanier et al. 1971

abilities (Stanier et al. 1971). In general lactic acid bacteria require sugars for growth, but the concentration varies with the species (Orla-Jensen 1942). They inhibit the growth of other bacteria by lowering the pH. Some strains can also produce antibiotics such as bacteriocins (Upreti and Hinsdill 1973). Others can produce hydrogen peroxide in the presence of oxygen and this may also contribute to their anti-microbial activity (Gilliland and Speck 1975).

Another characteristic of lactic acid bacteria is their salt tolerance. The lower salt concentrations, between 1.8-2.25%, favour the growth of heterofermenters, while higher concentrations favour the homofermenters (Pederson and Albury 1969). Temperature also affects the growth of the lactic acid bacteria. The majority of the lactobacilli are mesophiles with an optimum growth temperature in the range 25-35°C.

LACTOBACILLIC FERMENTATIONS

The process of fermentation has many advantages for man. Besides helping to preserve food, it contributes to the flavour, bouquet and texture and makes available more varieties of food. It also increases the range of raw materials available for food by detoxifying and predigesting otherwise useless plant parts. It can generate new food components such as vitamins and essential amino acids not present in the original raw material, thus improving its nutritional value. Finally, although the fermented food itself may not supply the additional nutrient directly, it may act by encouraging a beneficial gut microflora, or assist in the ingestion of other food components which are not themselves fermented prior to consumption (Stanton and Yeoh 1978).

The lactobacilli are able to carry out their metabolic activities with little change of the substrate. Thus the food is not decomposed to its basic components and most of the nutritive value of the raw material is retained (Pederson 1971). A wide variety of fermentation processes are carried out by these bacteria, ranging from the more simple ones, in which only one species is responsible to those in which there is a succession of microorganisms.

Table 3 gives some examples of fermentations which show that lactobacilli are involved in a variety of food fermentations. In many of them carbohydrates are either the main substrate fermented, or are added to the main substrate to supply the carbon source needed for the growth of the lactobacilli. The carbohydrate may be in the form of sugars or starches, i.e. starch assisted fermentations.

Examples of starch based fermented foods produced by lactobacillic fermentations are the cereal products such as idli, saké, sourdough, bantu beer, poi, busa, pulque and soya sauce. Soya sauce fermentation is a two stage process, the first being a solid mould fermentation step carried out by fungi of the genus Aspergillus such as A. oryzae or A. soyae. The second step is

Table 3. Some Lactobacillic Fermentations

Substrate	Microorganisms	Product
Cabbage	L. mesenteroides, L. brevis, L. plantarum	Sauerkraut
Beef	Homofermentative lactobacilli, P. cereviseae	Salami
Milk	S. thermophilus L. bulgaricus	Yoghurt
Skim milk	L. acidophilus	Acidophilus milk
Milk	S. lactis, L. casei	Cheddar cheese
Anchovies	Lactobacilli, other bacteria	Budu (Malaysia)

(continued)

Substrate	Microorganisms	Product
Acetes shrimps	Lactobacilli, other bacteria	Belacan (Malaysia)
Acetes shrimps	Lactobacilli, other bacteria	Cincaluk (Malaysia)
Fish	L. mesenteroides P. cereviseae L. plantarum S. faecalis Micrococci, yeasts	Burung dalag (Philippines)
Fish	Lactobacilli	Silage
Rice	L. mesenteroides S. faecalis P. cereviseae	Idli (India)
Rice	L. homohoichi L. heterohoichi Aspergillus spp. Saccharomyces	Saké (Japan)
Wheat	L. brevis, L. plantarum L. sanfrancisco Yeasts	Sourdough
Maize	L. delbrueckii L. bulgaricus	Bantu beer (Africa)
Taro	L. delbrueckii L. pastorianus L. brevis L. pentoaceticus S. lactis S. kefir Yeasts	Poi (Hawaii)
Millet	L. plantarum L. delbrueckii Saccharomyces busae asiaticae	Busa (Egypt)
Agave	Lactobacilli, L. mesenteroides Saccharomyces carabajali	Pulque (Mexico)
Soyabean	Aspergillus spp. L. delbrueckii P. soyae Saccharomyces rouxii	Soya sauce
Sucrose	L. mesenteroides	Dextran
Starch hydrolysate	L. delbrueckii	Lactic acid
Ginger flavoured sugar solution	Saccharomyces pyriformis L. vermiformis	Ginger beer

Sources: Frazier 1967; Stanton and Wallbridge 1969; Pederson 1971; Carr 1973; Stanton and Yeoh 1978.

the brine fermentation, and a variety of bacteria are active in this stage. The high salt content (18-20%) controls the microflora, and the microaerophilic conditions at the bottom of the vat favour particularly the growth of L. delbrueckii and P. soyae. These lactic acid bacteria are responsible for lowering the pH from about 6.7 to below 5.0, and are also responsible for contributing to the flavour (Yokotsuka 1960; Pederson 1971; Yeoh and Lee 1978; Merican 1978).

Other examples of starch assisted fermentations are the fermented fish products. The traditional method of preserving fish in the tropics is by fermentation and a variety of fish sauces and fish pastes are available. The anaerobic conditions for fermentation and the high salt content (20-25% in the end product) limit the microflora to few species only. Predominant amongst those that are able to grow are the lactobacilli. Since the carbohydrate content of the fish is low, carbohydrate sources such as cooked rice are added to achieve a more vigorous fermentation. An example is the burung dalag of the Philippines. It is made by mixing fish with rice and 10-12% of salt. Orillo and Pederson (1968) isolated strains of L. mesenteroides, P. cereviseae, L. plantarum, S. faecalis, as well as strains of Micrococcus and some yeasts. A similar product found in Malaysia is cincaluk, where the Acetes spp shrimps are mixed with cooked rice and salt and allowed to ferment.

This principle has been extended to the preservation of novel fermented products such as fish silage. There are two basic methods for producing fish silage, namely by direct addition of mineral or organic acids, or a mixture of the two, or by starch assisted fermentation methods. In the latter method, a carbohydrate source, viz. starch, is added to promote the fermentation. The lactobacilli ferment the carbohydrate to acids which lower the pH and prevent spoilage (Nilsson 1969).

In the biological process, the basic principle is a lactic acid fermentation, which sets in so rapidly and is so intensive that it extinguishes all other undesirable microbial and enzymatic activities (Nilsson and Rydin 1965). However, for a successful fermentation, relatively large quantities of carbohydrate, fermentable by lactic acid bacteria, must be available.

Biological Fish Silage

In our studies on biological fish silage, the method used
was as follows. Whole ungutted fish was minced using a bow
chopper/grinder for 2-3 minutes; a predetermined quantity of a
carbohydrate source, viz. cassava or sago was added. A source
of amylolytic enzymes, e.g. malt or ragi, and a starter culture
of lactic acid bacteria produced by a sauerkraut equivalent
fermentation was added. In some of the formulations salt was
also added to the mixture. The mixture was homogenised for a
further 2-3 minutes, then packed compactly into air-tight con-
tainers for the fermentation process. The fish was allowed to
ferment for 8-10 days at room temperature and, by the end of
this period, the pH had dropped from about 6.5 to below 5.0.

Two types of formulations have been found to be successful.
One is a fish/carbohydrate fermentation which involves the use
of 60% fish, 40% carbohydrate, 2% ragi and 2% starter culture.
However, it was found that if salt is added, the amount of
carbohydrate could be substantially reduced. The other formula
was therefore as follows; 80% fish, 15% carbohydrate, 5% ragi,
2% starter culture and 4% salt.

Figures 1 and 2 show the effect of the fish/carbohydrate
ratio on pH and lactic acid content respectively, while Figures
3 and 4 show the effects of salt on pH in a starch assisted
fermentation. Figures 1 and 3 show that, in a successful fermentation, there is a rapid drop in pH to below 5 with a corresponding increase in the lactic acid content (Figures 2 and 4).

Changes in the lactic acid bacteria population during fermentation have also been studied. Wirahadikusumah (1968) studied
the lactic acid bacteria predominant in biological fish silage.
He found that L. fermenti multiplied rapidly immediately after
ensiling and was responsible for the breakdown of starch to simple
sugars. The sugars were then rapidly fermented by L. brevis and
Streptococcus faecalis and the pH dropped rapidly to 4.5. Thus
he postulated that L. brevis and S. faecalis were responsible
for the reduction in pH.

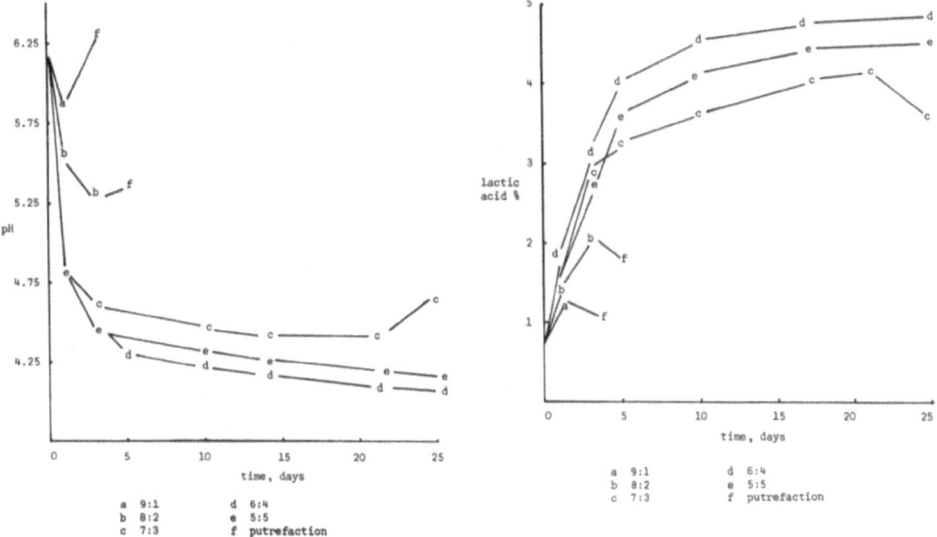

Figure 1. (left) Effect of Fish-Carbohydrate Ratio on pH

Figure 2. (right) Effect of Fish-Carbohydrate Ratio on Lactic Acid Production

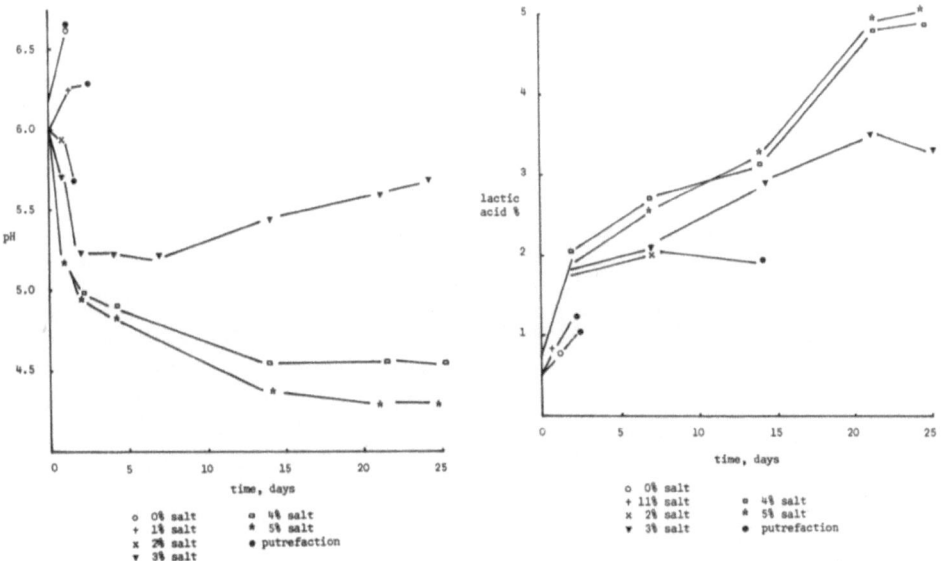

Figure 3. (left) Effect of Salt on pH in a Fish-Carbohydrate Silage

Figure 4. (right) Effect of Salt on Lactic Acid Production in a Fish-Carbohydrate Silage

A later study by Von Hofsten and Wirahadikusumah (1972) stated that the three predominant lactic acid bacteria in fish silage fermentation were L. mesenteroides, S. faecalis and L. plantarum. L. mesenteroides was dominant in the initial stage, but was succeeded by S. faecalis and L. plantarum in turn.

Figure 5 shows the changes in the lactic acid bacterial population of a silage prepared from 80% fish, 15% cassava, 5% ragi, 4% salt and 2% starter culture, carried out under ambient conditions. The initial pH was 6.10 and it dropped to 4.95 within 4 days, with a further drop to 4.80 after one week and remained constant for the rest of the trial period. The lactic acid bacteria were enumerated and classified according to the method of Pederson and Albury (1969).

The initial microflora showed a mixture of the various lactic acid bacteria reflecting the mixed population contained in the starter culture added. This was followed by the disappearance of L. mesenteroides, P. cereviseae, S. faecalis, L. brevis in

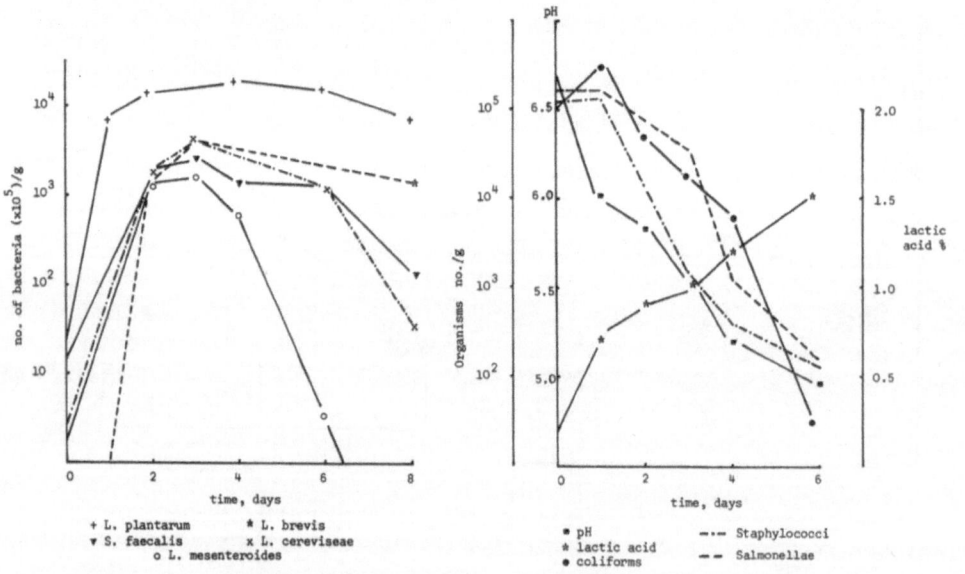

+ L. plantarum   * L. brevis
▼ S. faecalis   x L. cereviseae
o L. mesenteroides

● pH   --- Staphylococci
▲ lactic acid
● coliforms   -- Salmonellae

Figure 5.   Changes in the Lactic Acid Bacteria Population
(left)   in Fish Silage

Figure 6.   Changes in Fish Silage Inoculated with 0.5%
(right)   Salmonella and 0.5% Staphylococcus Cultures

that order; in the final stage of fermentation, only L. plantarum was present, as it is the most acid tolerant of the lactobacilli.

The survival of pathogens during fermentation was also monitored. Studies by Wirahadikusumah (1968) and other workers have shown that in good fish silage, no clostridia spp. or other pathogens such as Salmonella spp. are able to grow, or to produce toxins.

Figure 6 shows the changes occurring in a good fish silage which was initially inoculated with 0.5% Salmonella and 0.5% Staphylococcus cultures. It can be seen that the microbial counts for the three species of microorganisms monitored, viz. Salmonella, Staphylococcus and coliforms show a decline as the fermentation progressed. Thus the fermentation of fish by lactobacilli is effective in preventing hazards of microbial food poisoning, provided the conditions of fermentation are controlled. The fish silage as described is used for animal feed. However, experiments carried out have also shown that the method can be modified to produce a fermented fish product suitable for human consumption; this fish silage contains approximately 35% of protein in the final product (dry wt basis). It is therefore a good source of protein, unlike the traditional fermented fish products where the intake is limited by the high salt concentrations used in the process.

CONCLUSIONS

The examples described above, particularly the biological fish silage, illustrate the role of the lactobacilli in preserving raw materials in starch assisted food fermentations. The rapid increase in the world population has widened the gap between the available food supply and the demand. All methods of food preservation that can help to bridge this gap should be investigated. It is important to realise that fermentation can also play a major role in producing a wide variety of nutritious foods. In this respect, the contribution of starch assisted lactobacillic food fermentation methods should not be underestimated.

Acknowledgements: The author wishes to thank the Director

General of MARDI for his permission to present this paper and
Dr. A. Zaharudin Idrus, Head of the Agricultural Products
Utilization Division, MARDI for his kind interest and support.
The author is also indebted to Dr. W.R. Stanton for his advice
and encouragement in carrying out the studies.

REFERENCES

CARR, J.G. 1973. Lactics of the world unite, Lactic Acid Bacteria
in Beverages and Food, ed. J.G. Carr, C.V. Cutting and
G.C. Whiting; Academic Press, 1975: 369-80.

DOELLE, H.W. 1969. Bacterial Metabolism; Academic Press.

FRAZIER, W.C. 1967. Food Microbiology; (2nd ed.) TATA McGraw-
Hill Pub.

GILLILAND, S.E., and M.C. Speck 1975. Inhibition of psychro-
tropic bacteria by Lactobacilli and Pediococci in non-
fermented refrigerated foods, J. Food Sci. 40: 903.

INGRAM, M. 1973. The lactic acid bacteria - A broad view,
op. cit., ed. Carr et al.: 1-13.

INGRAM, M., and H. Luthi 1961. Microbiology of fruit juices,
Fruit and Vegetable Juice Processing Technology; AVI
Pub.: 117-63.

MERICAN, Z. 1978. Status of soy sauce research in Malaysia,
ASEAN Workshop on Soy Sauce Manufacturing Techniques;
Singapore.

NILSSON, R. 1969. A biological method in preventing the
deterioration of protein sources and products; lecture
in Brunswick (unpub.).

NILSSON, R., and C. Rydin 1965. A new method of ensiling
foodstuffs and feedstuffs of vegetable and animal origin,
Enzymologia 29(305): 126-42.

ORILLO, C.A., and C.S. Pederson 1968. Lactic acid bacterial
fermentation of 'burung dalag', App. Microbiol. 16(11):
1669-71.

ORLA-JENSEN, S. 1942. The Lactic Acid Bacteria; (2nd ed.)
I Kommission Hos Ejnar Munksgaard, Kobenhaven.

PEDERSON, C.S. 1971. Microbiology of Food Fermentations; AVI Pub.

PEDERSON, C.S., and M.N. Albury 1969. The Sauerkraut Fermentation; New York State Agric. Exp. Stat. bull. 824,
Cornell Univ., Geneva.

STANIER, R.Y., M. Doudoroff and E.A. Adelberg 1971. General
Microbiology; (3rd ed.) Macmillan Press.

STANTON, W.R., and A. Wallbridge 1969. Fermented food processes,
Proc. Biochem. 4(4): 45-51.

STANTON, W.R., and Q.L. Yeoh 1978. Microbial ecology applied to enhancing the safety of village fermented food processes, Proc. Reg. Conf. Technology for Rural Development, Kuala Lumpur.

UPRETI, G.C., and R.D. Hinsdill 1973. Isolation and Characteristic of a bacteriocin from a homofermentative lactobacillus, Anti-microbial Agents and Chemotheraphy, Oct. 47.

VON HOFSTEN, B., and S. Wirahadikusumah 1972. Preservation of fish and other protein rich products by lactic acid fermentation, Waste Recovery by Microorganisms, comp. W.R. Stanton, UNESCO/ICRO Workstudy; Dewan Bahasa dan Pustaka, Kuala Lumpur: 63-72.

WIRAHADIKUSUMAH, S. 1968. Preventing Clostridium botulinum Type E poisoning and fat rancidity by silage fermentation, Lantbrukshogsskolans Annaler 34: 551-689.

YEOH, Q.L., and G.C. Lee 1978. Halophilic lactic acid bacteria from soy brine mash (to be pub.).

YOKOTSUKA, T. 1960. Aroma and flavour of Japanese soy sauce, Adv. Food Res. 10: 75-134.

List of Authors and Participants

Puan Z.C. ALANG
Biology Department
Universiti Pertanian Malaysia
Serdang
MALAYSIA

Professor J. BARRAU
Sous-Directeur
Laboratoire d'Ethnobotanique
Museum National d'Histoire Naturelle
57, rue Cuvier
75005 Paris
FRANCE

Dr. R.H.V. CORLEY
Research Officer
Clonal Palm Oil Research Unit
Unipamol Malaysia Sdn. Bhd.
P.O. Box 101, Layang-Layang, Johor
MALAYSIA

Professor M. FLACH
Department of Tropical Crop Science
University of Agriculture
Ritzema Bosweg 32
Wageningen
THE NETHERLANDS

Dr. E.B. HOLMES
Agronomist
Department of Primary Industries
Kuk Agricultural Research Station
P.O. Box 339, Mt. Hagen
PAPUA NEW GUINEA

Professor F.H.C. KELLY
Professor of Industrial Chemistry
School of Applied Sciences
Universiti Sains Malaysia
Minden, Penang
MALAYSIA

Mr. J.H. BAKER
Technical Service Manager
Novo Industries
P.O. Box 2480, Kuala Lumpur
MALAYSIA

Mr. BOK Kim Wa
Production Manager
Stamford Chemical Industries Sdn. Bhd.
112, Jalan Semagat
P.O. Box 56, Petaling Jaya
MALAYSIA

Mr. FAH Ah Ngau
Kumpulan Guthrie Sdn. Bhd.
P.O. Box 2516, Kuala Lumpur
MALAYSIA

Dr. S. HARTO
Badan Pengajian & Penerapan Teknologi
Bogor Agricultural University
Fatemeta-IPB
Bogor
INDONESIA

Mr. HOO Ah Teng
Technical Professional Officer
M.I.D.A.
P.O. Box 618, Kuala Lumpur
MALAYSIA

Dr. LIE Goan-Hong
Nutrition Unit Diponegoro
National Institute for Medical
    Research & Development
Ministry of Health
c/o Seameo Tropmed - U.I. Building
Salemba 4, Jakarta
INDONESIA

Mr. LOONG Sing Guan
Agronomist, Agronomic Advisory Unit
Sime Darby Plantations
P.O. Box 202, Batu Tiga
Selangor
MALAYSIA

Mr. S. MATSUMOTO
Managing Director
Ajinomoto (M) Bhd.
P.O. Box 2507, Kuala Lumpur
MALAYSIA

Encik MUSTAPHA bin Haron
Principal Assistant Secretary
Ministry of Agriculture
Jalan Swettenham
Kuala Lumpur
MALAYSIA

Encik OMAR bin Abdul Razak
Ketua, Cawangan Hasil Tanaman
APU, MARDI
P.O. Box 202, Serdang
MALAYSIA

Dr. M. RAIMBAULT
Office de la Récherche Scientifique
et Technique Outre-Mer
ORSTOM Laboratory of Microbiology
IRCHA Research Center
91710 Vert le Petit
FRANCE

Dr. T. SATO
Technical Adviser
Association for International
Cooperation on Agriculture & Forestry
19 Ichibanchyo, Chiyoda-ku
Tokyo
JAPAN

Ir. R. SOERJONO
Director, Forest Research Institute
Agricultural Research & Development
   Agency
Department of Agriculture
P.O. Box 66, Bogor
INDONESIA

Dr. M.M. LUDLOW
Principal Research Scientist
C.S.I.R.O.
Division of Tropical Crops & Pastures
Cunningham Laboratory
Mill Road, St. Lucia, Qld. 4067
AUSTRALIA

Cik MONAIDA bt. Mohid
Department of Agriculture
Ministry of Agriculture
Jalan Swettenham
Kuala Lumpur
MALAYSIA

Dr. K. NEWCOMBE
Energy Planner
Office of Minerals & Energy
P.O. Box 2352, Konedobu
PAPUA NEW GUINEA

Dr. K. PAIJMANS
Institute of Earth Resources
CSIRO, Division of Land Use Research
P.O. Box 1666
Canberra City, A.C.T. 2601
AUSTRALIA

Encik S. RASYAD
Badan Pengajian & Penerapan Teknologi
Bogor Agricultural University
Fatemeta-IPB
Bogor
INDONESIA

Professor G. SIRÉN
Sveriges Landbruksuniversitet
Institutionen för Ekologi och
   Miljövård
S-75007 Uppsala
SWEDEN

Dr. W.R. STANTON
Consultant, Kemikro Sdn. Bhd.
Petaling P.O. Box 46
Old Klang Road
Kuala Lumpur
MALAYSIA

Encik SUBARI bin Shibani
Research Officer
APU, MARDI
P.O. Box 202, Serdang
MALAYSIA

Dr. T.Y. TAKAMURA
Associate Professor
Subtropical Plant Institute
Kyoto University
Sue, Kushimoto cho
Wakayama-Prefechue
JAPAN

Dr. TAN Koonlin
Stiftung Volkswagenwerk Research
Fellow 1978/79
Institute of Southeast Asian Studies
SINGAPORE
c/o Petaling P.O. Box 46
Old Klang Road
Kuala Lumpur
MALAYSIA

Dr. H. TAZUKE
Director, Ajinomoto (M) Sdn. Bhd.
P.O. Nox 2507, Kuala Lumpur
MALAYSIA

Mr. TEO Choo Kian
Department of Agriculture
Jabatan Pertanian Negeri Johor
P.O. Box 44, Johor Bahru
MALAYSIA

Mr. William TEOH Beng Suang
Managing Director
Dindings Tapioca Industry Sdn. Bhd.
II Tingkat Taman Ipoh Lima
Ipoh Garden South
Ipoh, Perak
MALAYSIA

Mr. S. WIJANDI
Dean, Badan Pengajian &
    Penerapan Teknologi
Bogor Agricultural University
Fatemeta-IPB
Bogor
INDONESIA

Mr. B.J. WOOD
Research & Development Controller
Sime Darby Plantations
A.A.V., P.O. Box 202
Batu Tiga, Selangor
MALAYSIA

Ms. YEOH Quee Lan
Research Officer, APU, MARDI
P.O. Box 202, Serdang
MALAYSIA

Cik ZAWIAH bt. Hashim
Research Officer, APU, MARDI
P.O. Box 202, Serdang
MALAYSIA